Phthalocyanines

Properties and
Applications

Volume 2

Phthalocyanines

Properties and

Applications

Volume 2

Edited by

C. C. Leznoff and A. B. P. Lever

C. C. Leznoff
Department of Chemistry
York University
4700 Keele Street
Downsview, Ontario
Canada   M3J1P3

A. B. P. Lever
Consulting & Editing Services, Ltd.
455 Sentinel Road #1010
Downsview, Ontario
Canada   M3J1V5

Library of Congress Cataloging-in-Publication Data

CIP pending.

© 1993 VCH Publishers, Inc.

Printed in the United States of America

ISBN 1-56081-544-2 VCH Publishers
ISBN 3-527-89544-2 VCH Verlagsgesellschaft

Printing History:
10 9 8 7 6 5 4 3 2 1

Published jointly by

VCH Publishers, Inc.
220 East 23rd Street
New York, New York 10010-4606

VCH Verlagsgesellschaft mbH
P.O. Box 10 11 61
D-6940 Weinheim
Federal Republic of Germany

VCH Publishers (UK) Ltd.
8 Wellington Court
Cambridge CB1 1HZ
United Kingdom

Contents

Contributors

Claudio Ercolani and **Barbara Floris**, Dipartimento di Chimica, Universita' Degli Studi di Roma, "La Sapienza", Piazzale Aldo Moro, 5, Rome 00185, Italy.

M. Hanack, U. Keppeler, A. Lange, A. Hirsch, and **R. Dieing**, Institut für Organische Chemie II, Lehrstuhl für Organische Chemie II, Universität Tübingen, Auf der Morgenstelle 18, D-7400 Tübingen 1, F.R.G.

Nagao Kobayashi, Pharmaceutical Institute, Tohoku University, Aobayama, Sendai 980, Japan.

Tetsuo Saji, Department of Chemical Engineering, Tokyo Institute of Technology, O-okayama, Meguro-ku, Tokyo 152, Japan.

Kenji Hanabusa and **Hirofusa Shirai**, Department of Functional Polymer Science, Faculty of Textile Science & Technology, Shinshu University, Ueda, 386 Japan

Jacques Simon and **Pierre Bassoul**, GRIMM, ESPCI-CNRS, 10 rue Vauquelin, 75231 Paris Cedex 05, France

Series Preface

Since their synthesis early this century, phthalocyanines have established themselves as blue and green dyestuffs par excellence. They are an important industrial commodity (output 45,000 tons in 1987) used primarily in inks (especially ballpoint pens), coloring for plastics and metal surfaces, and dyestuffs for jeans and other clothing. More recently their use as the photoconducting agent in photocopying machines heralds a resurgence of interest in these species. In the coming decade, their commercial utility is expected to have significant ramifications. Thus future potential uses of metal phthalocyanines, currently under study, include (I) sensing elements in chemical sensors, (II) electrochromic display devices, (III) photodynamic reagents for cancer therapy and other medical applications, (IV) applications to optical computer read/write discs, and related information storage systems, (V) catalysts for control of sulfur effluents, (VI) electrocatalysis for fuel cell applications, (VII) photovoltaic cell elements for energy generation, (VIII) laser dyes, (IX) new red-sensitive photocopying applications, (X) liquid crystal color display applications, and (XI) molecular metals and conducting polymers.

In recent years there has been a growth in the number of laboratories exploring the fundamental academic aspects of phthalocyanine chemistry. Interest has been focused, inter alia, on the synthesis of new types of soluble and unsymmetrical phthalocyanines, on the development of new approaches to the synthesis of polynuclear, bridged, and polymeric species, on their electronic structure and redox properties, and their electro- and photocatalytic reactivity.

It is timely therefore to consolidate these academic and applied aspects of phthalocyanine chemistry into a series of monographs that will present our current knowledge of this fascinating area of chemistry. In this series of monographs, we plan to bring together for the first time detailed critical coverage of the whole field of phthalocyanine chemistry in a sequence of chapters that should prove stimulating for industrial, medical, and academic researchers.

C. C. Leznoff
A. B. P. Lever

Preface

Volume 2, of a continuing Series, brings six Chapters continuing the coverage of phthalocyanine chemistry begun in Volume 1. The fascinating chemistry of single atom bridged dinuclear phthalocyanines is discussed by Ercolani and Floris (Chapter 1). Mössbauer spectra is a powerful technique for studying iron phthalocyanine complexes, especially the many polymeric complexes (Hanack, Keppeler, Lange, Hirsch and Dieing, Chapter 2). There are many other macrocyclic complexes based upon the phthalocyanine sub-unit but with interesting structural variations - these are covered by Kobayashi (Chapter 3).

Of the many properties of phthalocyanine species, their ability to form monolayers, multilayers or films is likely to prove of significant importance in the development of microelectronic devices, sensors etc. This topic is covered by Saji (Chapter 4) while the related area of liquid crystals, also of potential importance for electronic devices, is covered by Simon and Bassoul (Chapter 6). These two Chapters continue the coverage of this area, begun by Snow and Barger in Volume 1. The Chapter by Wöhrle in Volume 1 introduced the field of polymeric phthalocyanines and this is further developed from the catalytic viewpoint, in this Volume, by Hanabusa and Shirai (Chapter 5).

While Volume 1 was typeset, the fiscal realities of the early nineties required this Volume to be produced camera ready. The Editors have tried to maintain a common format from chapter to chapter but inevitably, with copy being generated in many countries on different word processors and printers, variations do occur. We hope that the small differences in appearance between chapters will not interfere with the enjoyment of this contribution.

C.C. Leznoff
A.B.P. Lever
July 1992

1

Metal

Phthalocyanine

Single-Atom

Bridged Dimers

Claudio Ercolani and
Barbara Floris

List of Abbreviations

AcOH	= acetic acid
DMA	= N, N-dimethylacetamide
DMF	= N, N-dimethyaformamide
DMSO	= dimethyl sulfoxide
im	= imidazole
meim	= methylimidazole
mepy	= methylpyridine
Np	= naphthaline
Pc	= phthalocyaninate anion
Ph	= phenyl
pip	= piperidine
PNP	= bis (triphenylphosphine) iminium cation
py	= pyridine
TBA	= tetrabutylammonium cation
TDPc	= tetradodecylphthalocyaninate anion
THF	= tetrahydrofuran
TOPc	= tetraoctylphthalocyaninate anion
TPP	= tetraphenylporphyrinate anion
TTBPc	= tetra (tert-butyl) phthalocyaninate anion

A. INTRODUCTION

The purpose of this chapter is the description of the synthesis and chemical-physical characterization of dimeric systems in which two metal phthalocyanine units (PcM) are held together by light atoms to give PcM-X-MPc bridged species (X = O, N, C). These complexes – as well as derivatives therefrom having the same M-X-M skeleton – can be classified as "single-atom bridged metal phthalocyanine dimers". Reference will also be made here to a few bridged dimers containing substituted phthalocyanine ligands and to a μ-oxo heterobimetallic mixed ligand porphyrinato-phthalocyaninato complex. μ-Oxo bridged polymeric species of general formula (PcM-O)$_n$, briefly mentioned in Volume 1 of this series (Chapter 2) and, more extensively, in this volume, will not be considered here.

The μ-oxo phthalocyanine dimers, which are characterized by the presence of the M-O-M bond system, fall within the large class of μ-oxo bimetallic complexes fairly extensively described in the literature [1-3]. Amongst these, the μ-oxo iron porphyrin dimers are important representative examples worthy of mention, since they are the final products irreversibly formed in the interaction of Fe(II)porphyrins with molecular oxygen, as in Scheme I [4].

The μ-nitrido and the μ-carbido phthalocyanine dimers, carrying the M-N-M and the M-C-M bond systems, respectively, are still quite rare species, which find counterparts only in a restricted number of μ-nitrido [5,6] and μ-carbido [7,8] iron porphyrin complexes. Several reasons, which include dioxygen activation, magnetic and electronic exchange in discrete systems, formation of mixed valence dimers, and high valency states for iron, have stimulated the synthesis and investigation of this class of bridged molecules. A strong limitation to an extensive and deep examination of such species is

$$(P)Fe(II) + O_2 \longrightarrow (P)Fe(III)(O_2^-)$$

$$(P)Fe(III)(O_2^-) + (P)Fe(II) \longrightarrow (P)Fe(III)\text{-}O\text{-}O\text{-}Fe(III)(P)$$

$$(P)Fe(III)\text{-}O\text{-}O\text{-}Fe(III)(P) \longrightarrow 2\ (P)Fe(IV)O$$

$$(P)Fe(IV)O + (P)Fe(II) \longrightarrow (P)Fe(III)\text{-}O\text{-}Fe(III)(P)$$

Scheme I

often represented by their low solubility in donor or nondonor solvents, with an implied great difficulty in their accurate purification and associated unequivocal identification. Thus, in order to avoid the erroneous formulation of the complexes, effort has to be made in using different techniques, including,

Table 1   Five-Coordinate Single-Atom Bridged Dimers[a]

$(PcAl)_2O$	**(1)**	$[(PcTi)_2O]^{2+}$	**(2)**
$(PcMn)_2O$	**(3)**	$(PcFe)_2O$	**(4)**[b]
$(PcFe)_2N$	**(5)**	$(PcFe)_2C$	**(6)**
$[(PcFe)_2N]^+$	**(7)**[c]	$(PcRu)_2N$	**(8)**
$(TTBPcFe)_2O$	**(9)**	$(TOPcFe)_2O$	**(10)**
$(TDPcFe)_2O$	**(11)**	$(TPP)Cr\text{-}O\text{-}FePc$	**(12)**

Six-Coordinate Single-Atom Bridged Dimers[a]

$[(X)PcMn]_2O$	(X = py, pip, 3-mepy, 4-mepy, 1-meim)	**(13-17)**
$[(X)PcFe]_2O$	(X = py, 4-mepy, 1-meim, pip)	**(18-21)**
$[(X)PcFe]N$	(X = Br)	**(22)**
$[\{(X)PcFe\}_2N]^+$	(X = py, 4-mepy, 1-meim, pip)	**(23-26)**
$[(X)PcFe]_2C$	(X = py, 4-mepy, 1-meim, pip; py, THF)	**(27-32)**
$[\{(X)Pc(-1)Fe\}_2N]^+$	(X = Br, CF$_3$CO$_2$, NO$_3$)	**(33-35)**
$[(X)Pc(-1)Fe]_2C$	(X = Cl, Br, NO$_3$)	**(36-38)**
$[\{(X)PcFe\}_2C]^{2-}$	(X = CN, OCN, SCN, F)	**(39-42)**

[a] The species reported are those effectively isolated in the solid state. Other species identified electrochemically (see the following) are not included.
[b] μ-Oxo(1) (**4a**) and μ-Oxo(2) (**4b**).
[c] Counterion PF_6^- (**7a**) or I_3^- (**7b**).

whenever possible, single-crystal x-ray work, for definition of their molecular and electronic structure.

Table 1 summarizes the various species examined in the present review.

B. FIVE-COORDINATE DIMERS

The dimers dealt with here are Al, Ti, Mn, Fe, and Ru derivatives. A lanthanide μ-oxo species, (PcLa)$_2$O, has been mentioned, but without characterization or properties [9].

i. (PcAl)$_2$O

Single-crystal x-ray work has allowed the elucidation of the structure of the only known μ-oxo dimer, (PcAl)$_2$O (1), having a nontransition element as the central metal atom [10]. The structure (Figure 1) locates the O atom in the center of symmetry of the molecule, to which the two (PcAl) units are bound, through Al, on opposite sites (Al-O, 1.679 Å; Al-O-Al, 180°). Each (PcAl) unit forms a square pyramid with the four inner N atoms at the base, and the pentacoordinate Al atom in the apical position, 0.459 Å above the center of the N$_4$ chromophore. An interesting feature of the structure is the eclipsed position of the two cofacial Pc rings (4.27 Å apart), allowed by the out-of-plane position of the Al atoms. Parallel single-crystal x-ray work on the polymeric μ-F species (PcGaF)$_n$ [10], in which the Ga atom is coplanar with

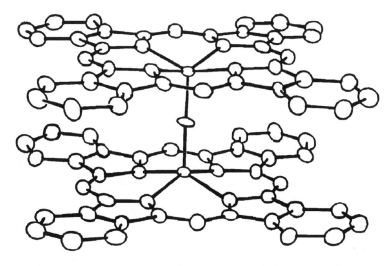

Figure 1 X-ray structure of (PcAl)$_2$O. (Reproduced from [10], with permission.)

the N_4 coordinating system within each macrocyclic ring, has shown that an eclipsed arrangement of two adjacent Pc units is still possible with an interplanar distance of 3.87 Å. Since $(PcAl)_2O$ is the only one single-atom bridged dimer of known structure among the complexes of formula $(PcM)_2X$, no comparison is currently possible with similar or alternative (staggered) cofacial arrangements. It is true, however, that when the interplanar distance is *ca.* 3.5 Å or lower, as occurs in μ-oxo polymeric systems of the type $(PcMO)_n$ (see refs. cited in [10]), in partially oxidized systems therefrom ([11] and refs. therein), or other related electrically highly conducting (PcM) materials [12-17], the staggered relative orientation of adjacent Pc rings is normally observed. It is anticipated that a staggered arrangement is also found in Mn and Fe μ-X dimers discussed here (see the following). IR, Raman, and surface-enhanced Raman spectroscopic (SERS) studies [18] have allowed identification of the $\nu_{as}(Al-O-Al)$, located at 1050 cm^{-1} (Table 2). Films of this μ-oxo species have also been examined by transmission electron microscopy in an attempt to correlate the photoactivity of organic semiconductors and the degree of crystallinity of the materials [19(a)]. The photoelectrochemistry of 1 deposited as thin films on SnO_2 has been also examined for the estimation of the degree of energy conversion under white light illumination and in contact with an acidic solution containing the I_3^-/I^- system. The value of J_{sc} at 35 mW cm^{-2} is 245 μA cm^{-2} [19(b)].

ii. $[(PcTi)_2O]^{2+}$

A short description of the synthesis of this cationic Ti(IV) dimer, isolated as a perchlorate salt $[(PcTi)_2O)](ClO_4)_2$, has been reported [20]. It is prepared by suspending PcTiO in CH_2Cl_2/CH_3CN in the presence of $HClO_4$. The complex shows an IR absorption at 820 cm^{-1}, tentatively assigned to the antisymmetric Ti-O-Ti stretch (Table 2). In the presence of amines the Ti(IV) dimer regenerates PcTiO.

iii. $(PcMn)_2O$

$(PcMn)_2O$ (3) and some of its N-base adducts of formula [(N-base)PcMn]$_2$O (13-15, 17) are the oldest single-atom bridged metal phthalocyanine dimers reported in the literature [21-27]. This investigation was developed in connection with the involvement of manganese in processes in which water is photo-oxidized to dioxygen [21, 25, 26]. Thus, the reaction of PcMn with O_2 in pyridine leads to the formation of the dimeric species $[(py)PcMn(III)]_2O$ (13), originally proposed as Mn(IV) monomer [21]. Heating of the adduct 13 in vacuo ($100°C$, 10^{-2} mm Hg) leads to 3.

It has been suggested that the Mn-O-Mn system in 3 is linear or quasi-linear [28], as found by x-ray determination of 13 [24], since the elimination of pyridine from it should not change significantly the geometry of the oxygen-bridged bimetallic system. More difficult is the approach to the definition of the ground and spin state of Mn(III) in **3** since a detailed magnetic investigation on this species is not available. Infrared spectra do not identify unequivocally any absorption due to the Mn-O-Mn system [25] (Table 2).

The reaction of PcMn with O_2 in pyridine is reversible and develops through the formation of intermediates, **13** being the ultimate product formed [21-23, 25, 26]. Such a reaction is strongly affected by the purity of the solvent and the presence of additional reactants. Indeed, no reaction occurs between PcMn and O_2 if pyridine is not freed from acids or when the solvent is rigorously purified and dried. A visible spectral examination confirms that the process can be completely reversed by prolonged N_2 bubbling or by heating the pyridine solution [26]. Evidence has been produced that an easily detectable intermediate, having its maximum visible absorption at 710 nm, is the $PcMn(O_2)$ dioxygen adduct [29], excluding either a peroxy-bridged PcMn dimer or the monomeric PcMnOH [26]. This latter can be obtained from a pyridine solution containing **13** (λ_{max} = 620 nm) by addition of water, followed by precipitation on addition of ether. Scheme II summarizes reactions and products of PcMn and O_2 in pyridine under a variety of conditions.

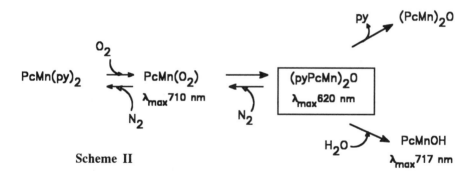

Scheme II

Reaction of PcMn with O_2 in N,N-dimethylacetamide (DMA) allows isolation and characterization of solid $PcMn(O_2)$, as well as of $(PcMn)_2O$ [26, 29]. Scheme III summarizes the various species formed in DMA under a variety of conditions.

Photoreduction of monomeric Mn(III) intermediates, obtained from **3**, followed by formation of PcMn, has been observed when **3** is dissolved in DMA or N,N-dimethylformamide (DMF) and the solution flashed with visible light [30]. PcMn is also formed from **3**, although with a low quantum yield, when the μ-oxo dimer is dissolved in pyridine in the presence of traces of

Table 2 IR and Other Physical Properties of μ-X Dimers.

Compound	$\nu_{as}(M\text{-}X\text{-}M)^a$ (cm^{-1})	λ_{max} (solvent) (nm)	$\mu_B{}^b$	Decomposition temperature,°C^c	Ref.
(PcAl)$_2$O (**1**)	1050(s)^d	642(film)			[18]
(PcTi)$_2$O (**2**)	820(m-s)				[20]
(PcMn)$_2$O (**3**)	not observed	620(py)			[26]
(PcFe)$_2$O (**4a**)	852(s),824(s)	695(solid, H$_2$SO$_4$)	2.09^e	330	[28, 43, 46]
	822	695 (1-ClNp)			[48]
(**4b**)	not observed	695(solid, H$_2$SO$_4$)	1.38-1.42^e		[28, 44]
(PcFe)$_2{}^{18}$O	806				[28, 43]
(PcFe)$_2$N (**5**)	915(vs)	625 (py)	2.13^f		[74, 75, 80]
		690 (solid)	2.33^f		[77]
[(PcFe)$_2$N]$^+$ (**7**)	absent	637			[76]
(PcFe)$_2$C (**6**)	990(s)	620 (py)		>300	[84, 85]
(PcRu)$_2$N (**9**)	1040(m-s)	600 (py)	1.8^f	>300	[87]
(PcRu)$_2{}^{15}$N	1023(m-s)				[87]
(TTBPcFe)$_2$O (**10**)	not identified	688 (toluene)			[96]
(TOPcFe)$_2$O (**11**)		636 (THF)	1.7^e		[30]
(TDPcFe)$_2$O (**12**)	810	633 (THF)	2.0^e		[30]
[(py)PcMn]$_2$O (**13**)	not observed	619 (py)	1.6-2.2^e	100^h	[23, 70]
[(pip)PcMn]$_2$O (**14**)	not observed	617 (pip)	wp^g		[25,70]
[(3-mepy)PcMn]$_2$O (**15**)	not observed	633 (3-mepy)	wp^g		[25]
[(4-mepy)PcMn]$_2$O (**16**)	not observed	621 (4-mepy)			[70]
[(1-meim)PcMn]$_2$O (**17**)	not observed	632 (1-meim)	wp^g		[70]
[(py)PcFe]$_2$O (**18**)	not observed	620 (py)	1.86^e	130-140^h	[70]
[(4-mepy)PcFe]$_2$O (**19**)	not observed	*ca.* 620 (4-mepy)	1.99^e	160-170^h	[70]
[(1-meim)PcFe]$_2$O (**20**)	not observed	*ca.* 620 (1-meim)	2.16^e	90-100,160-180^h	[70]
[(pip)PcFe]$_2$O (**21**)	not observed	*ca.* 620 (pip)	2.11^e	100-130^h	[70]
[(PcFe)$_2$C].Me$_2$CO	940				[85]
[(py)PcFe]$_2$C (**27**)	910				[85]
[(4-mepy)PcFe]$_2$C (**28**)	888				[85]
[(1-meim)PcFe]$_2$C (**29**)	940				[84, 85]
[(pip)PcFe]$_2$C (**30**)	910				[85]
[{(Br)PcFe}$_2$N]$^+$ (**31**)			2.34^f		[77]
[{(CF$_3$CO$_2$)PcFe}$_2$N]$^+$ (**32**)			1.98^f		[77]
PcFeOCr(TPP) (**12**)	not observed	431, 699 (1-ClNp/PhH)	3.12		[97]

^{a} Nujol mulls or KBr.

^{b} Room temperature magnetic moment in Bohr magnetons.

^{c} In vacuo or under nitrogen.

^{d} Film.

^{e} Per metal atom.

^{f} Per dimer.

^{g} Weak paramagnetism, probably due to contaminants or to strong spin coupling [25, 23].

^{h} Loss of N-base.

water. The cleavage of the Mn-O-Mn bridge probably passes through the formation of the intermediate PcMnOH. Also, **3** is photoreduced when dissolved in ethanol (10^{-5} *M*) in the presence of a base (10^{-2} *M*).

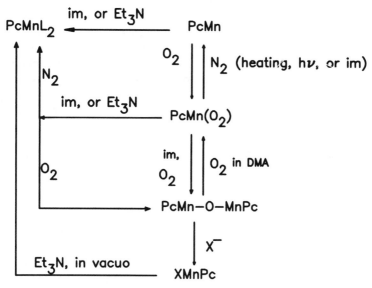

Scheme III Reactivity of PcMn in DMA

Electrochemical studies [27] show that the μ-oxo species **3** in pyridine solution undergoes: (1) a reversible one-electron oxidation at +0.35 V (*versus* SCE) to the mixed-valence cationic species $[PcMn(IV)-O-Mn(III)Pc]^+$ (the complex should be regarded as having pyridine coordinated externally at the Mn centers). The same oxidation takes place irreversibly at +0.25 V in DMF; (2) an irreversible reduction at −0.8 V, requiring four electrons per dimeric unit, and generating a species, characterized by its electronic spectrum as $Pc(-3)Mn(II)(py)_2$, which can also be obtained by reduction of $PcMn(II)(py)_2$; reduction, then, implies the cleavage of the dimer, thus accounting for the irreversibility of the electrochemical reduction process.

iv. (PcFe)₂O

Synthesis and structure. A detailed description of the synthetic aspects and properties [31-37] and structural characterization [38-40] of a large number of metal phthalocyanines, including those carrying metal ions of the first transition series, was given in the 1930s. Since then, the molecular structure and behavior of PcFe, one of the most representative examples of a synthetic analog of the porphyrin core in hemeproteins and Cytochrome P450, have been widely investigated. It is surprising, however, that it took more than forty years before it could be realized that this macrocyclic porphyrin-like complex, indefinitely stable in air as a solid, can interact at room temperature with molecular oxygen in a variety of conditions, ultimately being quantitatively converted into the μ-oxo dimer $(PcFe)_2O$ [28, 41-44], thus closely resembling the behavior of Fe(II)-porphyrins (Scheme I). Indeed, in 1969 it was observed [45] that solutions of PcFe in dimethyl sulfoxide (DMSO) were unstable in contact with air, and an unspecified precipitate, likely to be $(PcFe)_2O$ [46], was slowly formed. Furthermore, the spectral behavior of solutions in DMSO of PcFe in air was interpreted as due to the formation of a dimer [47]. This was contested later, and reinterpreted as due to the formation of $(PcFe)_2O$ [46].

   $(PcFe)_2O$ is a unique example in the field of metal phthalocyanines, or other similar porphyrin or porphyrin-like systems, of a μ-oxo dimer that has been isolated and characterized in two quite distinct isomers having a bent [μ-Oxo(1), Figure 2(a)] and a linear Fe-O-Fe bond moiety [μ-Oxo(2), Figure 2(b)] [28, 43, 44, 48]. These two isomers are blue, microcrystalline, air-stable materials, which are thermally stable up to temperatures of 300-350°C (Table 2). They show different x-ray powder patterns, IR and Mössbauer spectral data, and different magnetic behavior. Moreover, they show quite similar reflectance spectra and identical visible solution spectra in 96% H_2SO_4 (Table 2). They also exhibit identical spectral and chemical changes when dissolved in N-bases (py, 4-mepy, etc.).

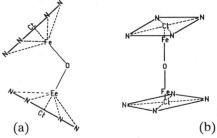

$$(a) \qquad\qquad\qquad (b)$$

Figure 2   (a), Structure of μ-Oxo(1);   (b), μ-Oxo(2). (Reproduced from [44], with permission.)

   μ-Oxo(1) is obtained from PcFe [28, 43, 44] or from $(PNP)[PcFe(OH)_2]$ [48]. Reaction of PcFe, suspended in DMF, DMA, dioxane, or tetrahydrofuran (THF), in the presence of O_2 or air, quantitatively produces μ-Oxo(1) [28, 43]; gas-volumetric measurements in DMF at room temperature show that the reaction proceeds as follows

$$4 \text{ PcFe } + \text{ } O_2 \longrightarrow 2 \text{ } (PcFe)_2O$$

μ-Oxo(1) from PcFe and O_2 has also been observed to form in 1-chloronaphthalene [49]. From (PNP)[PcFe(OH)$_2$] [48], (PcFe)$_2$O is also easily obtained, as μ-Oxo(1), in dichloromethane, chloroform, or acetone; the process of formation in dichloromethane is made faster by addition of AcOH or I_2.

The isomer μ-Oxo(2) is obtained from PcFe or from μ-Oxo(1) [28, 44]. It is best prepared by dissolution of the latter in 96% H_2SO_4, followed by reprecipitation in iced water [44]. Other methods of preparation are from μ-Oxo(1) in 1-chloronaphthalene and 1-butanamine [28] or by oxidation of PcFe in 96% H_2SO_4 [28, 44]. Formation of μ-Oxo(2) has also been observed in DMSO [28]. Either μ-Oxo(1) or μ-Oxo(2), or a mixture of them, is formed by suspending PcFe or PcFe(py)$_2$ (partially dissolved) in toluene under $p(O_2)$ = 60-70 atm [50].

IR, Raman, and structural data [28, 43, 44, 48] and magnetic susceptibility and Mössbauer measurements [28, 43, 44, 51] have been mainly used for the identification of the molecular and electronic structure of both μ-Oxo(1) and μ-Oxo(2). This chemical physical investigation has clearly shown that samples of each isomer often contain small amounts of the other isomer. Magnetic susceptibility, EPR, and Mössbauer measurements have also frequently indicated the presence of Fe(III) high-spin monomeric contaminants, or phthalocyanine radicals, as is normal with PcFe derivatives in general.

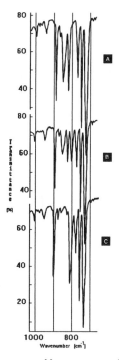

Figure 3 IR spectrum of μ-Oxo(1). (A) ^{16}O, (B) 70% ^{18}O, (C) 99% ^{18}O. (Reproduced from [28], with permission.)

The IR spectrum of μ-Oxo(1) shows medium-to-strong absorptions at 852 and 824 cm^{-1} [28, 43, 48], assigned as ν_{as}(Fe-O-Fe) on the basis of ^{18}O labeling [28, 43] (Table 2). Figure 3 shows the IR spectra of μ-Oxo(1) containing ^{16}O [Figure 3(A)], and enriched with 70% and 99% ^{18}O [Figure 3(B) and 3(C)]. Figure 3(C) shows that the presence in μ-Oxo(1) of 99% ^{18}O determines the complete disappearance of the bands at 852 and 824 cm^{-1} and the appearance of a strong absorption at lower energy (806 cm^{-1}), in line with expectation. Intermediate changes are produced by 70% ^{18}O labeling. These spectral variations appear to be in contrast with the expected changes for an alternative μ-dioxo formulation [28, 43], as proposed elsewhere [52, 53]. μ-Oxo(2) does not show bands attributable to the Fe-O-Fe moiety.

Mössbauer spectral data and magnetic susceptibility measurements establish for both isomers the presence of antiferromagnetically interacting couples of high-spin (5/2, 5/2) Fe(III) [28, 44, 51]. The Mössbauer spectra of μ-Oxo(1) (Figure 4) and μ-Oxo(2) (Figure 5) give isomer shift values of 0.36 and 0.26 mm s^{-1}, respectively, consistent with the presence of Fe(III) (Table 3). Antiferromagnetic coupling is proved by the decrease of the magnetic moment as a function of the temperature in the range 300-4 K. Good fitting is observed for interacting high-spin Fe(III) couples (5/2, 5/2), as shown in Figures 6 and 7. The coupling constant J is -120 and -195 cm^{-1} for μ-Oxo(1) and μ-Oxo(2), respectively. The much higher J value found for μ-Oxo(2) is in

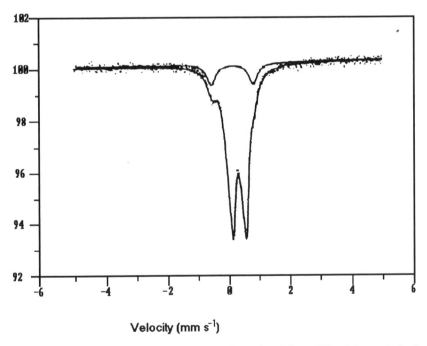

Velocity (mm s^{-1})

Figure 4 Mössbauer spectrum of μ-Oxo(1). (Reproduced from [51], with permission.)

Table 3 Mössbauer Data for μ-oxo, μ-nitrido, and μ-carbido Species (for this set of data, see also Figure 14).

Complex	T (K)	δ^a mm s^{-1}	ΔE_Q mm s^{-1}	Γ mm s^{-1}	Refs.
Five-coordinate PcFe(III) μ-oxo dimers					
(PcFe)$_2$O [μ-Oxo(1)]	77	0.36	0.44	0.16	[51]
(PcFe)$_2$O [μ-Oxo(2)]	77	0.26	1.26	0.14	[44]
Six-coordinate PcFe(III) μ-oxo dimers					
[(4-mepy)PcFe]$_2$O	4.2	0.20	1.76	0.15	[70]
[(py)PcFe]$_2$O	4.2	0.18	1.73	0.15	[70]
[(pip)PcFe]$_2$O	4.2	0.19	1.61	0.16	[70]
[(1-meim)PcFe]$_2$O	4.2	0.17	1.58	0.14	[70]
Five-coordinate PcFe(III$_{1/2}$) μ-nitrido dimers					
(PcFe)$_2$N	77	0.06	1.76	0.19	[80]
	4.2	0.06	1.78		[77]
Five-coordinate PcFe(IV) μ-nitrido species					
[(PcFe)$_2$N](PF$_6$)	77	-0.10	2.06	0.16	[80]
[(PcFe)$_2$N]I$_3$	77	-0.11	1.98	0.14	[83]
Six-coordinate PcFe(IV) μ-nitrido species					
[{(py)PcFe}$_2$N]I$_3$	77	-0.11	1.86	0.17	[83]
[(4-mepy)PcFe]$_2$N(PF$_6$)	77	-0.10	1.76	0.17	[80]
[(py)PcFe]$_2$N(PF$_6$)	77	-0.09	1.76	0.20	[80]
[(pip)PcFe]$_2$N(PF$_6$)	77	-0.09	1.73	0.16	[80]
[(1-meim)PcFe]$_2$N(PF$_6$)	77	-0.09	1.52	0.16	[80]
Six-coordinate Pc(-1)$^\cdot$ Fe(IV) species					
[{BrPc(-1)$_2$N]Br (65%)	4.2	-0.08	1.74		[77]
(35%)	4.2	-0.10	2.02		

(continued)

Table 3 Mössbauer Data for μ-oxo, μ-nitrido, and μ-carbido Species (for this set of data, see also Figure 14) (*continued*).

Complex	T (K)	δ^a mm s^{-1}	ΔE_Q mm s^{-1}	Γ mm s^{-1}	Refs.
		Six-coordinate Pc(−1)˙ Fe(IV) species			
[{(CF₃CO₂)PcFe(−1)}₂N](CF₃CO₂)	4.2	−0.10	1.82		[77]
		Five-coordinate PcFe(IV) μ-carbido species			
(PcFe)₂C	77	−0.16	2.69	0.11	[85]
	77	−0.04	2.67	0.16	[83]
		Six-coordinate PcFe(IV) μ-carbido species			
[{(2Me₂CO)PcFe}₂C]	77	−0.10	1.46	0.15	[85]
[(4-mepy)PcFe]₂C	77	0.03	1.19	0.13	[85]
[(py)PcFe]₂C	77	0.01	1.16	0.13	[85]
[{(py)PcFe}₂C].0.5py	77	0.01	1.14	0.13	[83]
[(pip)PcFe]₂C	77	0.01	1.11	0.13	[85]
[(1-meim)PcFe]₂C	77	0.01	0.94	0.12	[85]
[{(THF)PcFe}₂C].THF	77	0.00	1.84	0.16	[83]
(TBA)₂[{(F)PcFe}₂C].5H₂O	77	0.01	1.14	0.21	[83]
(TBA)₂[{(CN)PcFe}₂C].2.5CH₂Cl₂	77	0.01	0.22	0.17	[83]
(TBA)₂[{(OCN)PcFe}₂C].2H₂O					
(40.3%)	77	0.03	0.78	0.16	[83]
(59.7%)		0.03	1.18	0.18	
(TBA)₂[{(SCN)PcFe}₂C].2H₂O	77	0.03	1.18	0.18	[82]
		Six-coordinate Pc(−1)˙ Fe(IV) μ-carbido species			
[(Cl)Pc(−1)Fe]₂C^b	77	0.00	1.16	0.22	[83]
[(Br)Pc(−1)Fe]₂C^b	77	0.00	1.65	0.23	[83]
[(NO₃)Pc(−1)Fe]₂C^b	77	0.00	1.28	0.23	[83]

^{a} Referred to the Fe metal
^{b} For the complete formulation of this species see [83].

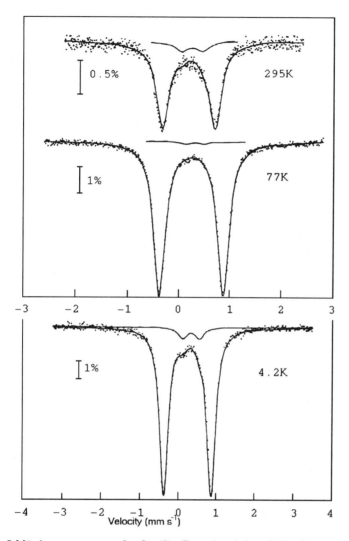

Figure 5 Mössbauer spectrum of μ-Oxo(2). (Reproduced from [44], with permission.)

keeping with expectation, since a linear Fe-O-Fe moiety [(b), Figure 2] should favor electronic exchange and magnetic coupling with respect to the bent system proposed for μ-Oxo(1) [Figure 2(a)]. In an IR investigation in the region 600-180 cm^{-1} a band at 280 cm^{-1} has been assigned for μ-Oxo(1) to the Fe-N$_{(Pc)}$ stretching motion by using $^{54/57}$Fe substitution [54].

(PcFe)$_2$O has been alternatively formulated as: (1) a μ-peroxo species, PcFe(III)-O-O-Fe(III)Pc [52, 53]; (2) an acidic Fe(II) species H$_2$[(PcFe)$_2$O] [55-57]; and (3) an Fe(II)-tetrasulfophthalocyanine [for μ-Oxo(2)] [58].

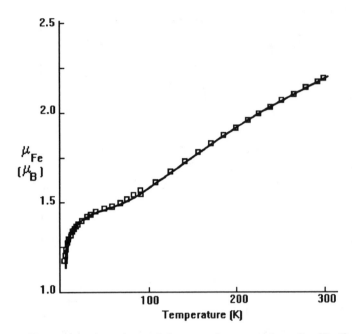

Figure 6 Temperature dependence of the magnetic moment for μ-Oxo(1). (Reproduced from [51], with permission.)

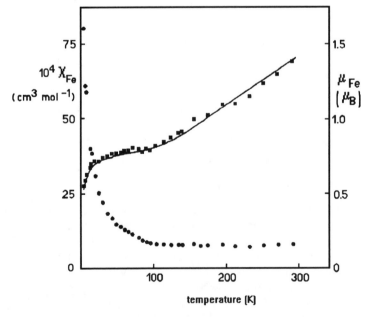

Figure 7 Temperature dependence of the magnetic moment and magnetic susceptibility for μ-Oxo(2). (Reproduced from [44], with permission.)

However, these formulations were not adequately supported and eventually proved wrong, as discussed below.

In case (1), a solid oxygen-containing species "S", formed by the reaction of PcFe with dioxygen in DMSO, has been briefly reported [52]. It has been given the μ-dioxo formula PcFe-O$_2$-FePc, mainly on the basis of elemental analysis and reference to previous kinetic work carried out on the same reaction in 96% H$_2$SO$_4$ [42], recently reconsidered [59]. Under "some preparative conditions", a crystalline modification of "S", named "B", has been irreproducibly obtained [52]. On the basis of the reported physical and chemical properties, there is no doubt that "S" (IR, μ_B, reaction with heterocyclic N-bases) and "B" (IR, x-ray powder diagram) coincide with μ-Oxo(1) and μ-Oxo(2), respectively. Later, the same author [53], in a discussion of mainly already published material, confirmed the occurrence of the two isomers "S" and "B" (renamed FePcoxg$_1$ and FePcoxg$_2$) and the presence of Fe(III) in both species, in contrast with other findings [58], and indicated new synthetic methods for the preparation of very small amounts of pure "B" [μ-Oxo(2)], free from the presence of "S" [μ-Oxo(1)], as impurity. However, Mössbauer spectra were not reported nor quantitatively discussed, thus not allowing an objective control of the effective purity. In the same contribution [53] it is uncertain whether the μ-oxo formulation, given elsewhere [28, 43, 44, 48, 51], is accepted, or the μ-dioxo formulation is still preferred for both "S" and "B". The same author has reported on an imidazole adduct of formula (im)PcFe-O-O-FePc(im) [60], most likely a μ-oxo dimer.

For (2), in an attempt to synthesize tetra-(*tert*-butyl)-phthalocyaninatoiron(II), TTBPcFe, an oxygen-containing species has been isolated, which, on the basis of elemental analyses, molecular weight measurements, IR and x-ray photoelectron spectra, and also of chemical properties, has been assigned the formula H$_2$(TTBPcFe)$_2$O , containing Fe(II). The same acidic formula has been automatically transferred to the unsubstituted μ-oxo dimer (PcFe)$_2$O by Russian authors [55]. A re-examination of some of the properties (mass spectra, IR and UV- visible measurements, thermal behavior) indicated as reasonable the formulation H$_2$[(TTBPcFe)$_2$O] [57]. Later this formula was reconfirmed by the same Russian authors [56], although in the meantime evidence [28, 43, 48] supporting the non acidic Fe(III) dimeric formula (PcFe)$_2$O had been published. The authors [56] confirmed the acidic formula on the basis of the rather peculiar reaction

$$2 \text{ TTBPcFe} + \text{H}_2\text{O} \longrightarrow \text{H}_2[(\text{TTBPcFe})_2\text{O}]$$

in which water is dissociated in 2H$^+$ and O^{2-}. The formula does not account for the thermal stability of the μ-oxo species, nor for the presence of Fe(III), proved by the Mössbauer spectra [44, 51]. Finally, the authors acknowledge that the reaction of PcFe with O$_2$ has been established "fairly reliably" [56, 61], but avoid drawing the correct conclusions.

In case (3), a Mössbauer investigation of the two isomers μ-Oxo(1) and μ-Oxo(2) [58], prepared as described previously [28] from PcFe and O_2 in DMF and 96% H_2SO_4, respectively, gave isomer shift and quadrupole splitting values confirming the presence of Fe(III) and its formula for μ-Oxo(1). For μ-Oxo(2), instead, the Mössbauer spectral features observed suggested the presence of low-spin Fe(II) in a sulfonated phthalocyanine species, identified as TSPcFe(II). This hypothesis has been argued [44, 51] on the basis of the following facts: (1) a sulfonation reaction of the phthalocyanine ring cannot occur in 96% H_2SO_4; (2) TSPcFe(II) is not stable in presence of molecular oxygen in H_2SO_4 and thus cannot be isolated therefrom; (3) TSPcFe(II) is soluble in H_2SO_4 and, therefore, cannot be isolated from that solvent by precipitation; (4) no bands of SO_3H are present in the expected region of the IR spectrum reported.

Apart from $PcFeCl_2$ [62], which, however, contains the π-radical cation $(Pc)^{\cdot +}$, $(PcFe)_2O$ has been cited [28, 51] as the first well-authenticated iron phthalocyanine derivative containing Fe in an oxidation state higher than two. It should be emphasized, however, that in recent times, concomitant with the isolation and characterization of $(PcFe)_2O$, convincing information has been reported for the presence of Fe(III) in the complex formulated as PcFeCl [48, 63, 64], alternatively presented as a hydrochloride of PcFe(II), PcFe.HCl (see, for instance, [55, 65]), but correctly formulated since 1938 [37]. Recently, the chemistry of monomeric Fe(III)-phthalocyanine complexes has been consistently enriched [66-68]. Indeed, Fe-phthalocyanine complexes having the central metal in formal oxidation states even higher than 3 [Fe(III$_{1/2}$), Fe(IV)] have also been reported (see the following).

The mechanism of formation. An accurate kinetic investigation on the oxygen uptake by PcFe has been carried out, both in DMSO [46] and in 96% H_2SO_4 [42, 59], following early qualitative results [41]. The experimental picture in both media is consistent with a multistep process, in which $(PcFe)_2O$ is the ultimate product formed. Initially, the interaction between O_2 and PcFe (or, better, a derivative with axially coordinated DMSO or HSO_4^- ions) forms, reversibly, $PcFe(O_2)$, which slowly transforms into $(PcFe)_2O$ (Scheme IV, in which R = reducing agent, likely to be the solvent or any related species).

A series of reactions takes place before the irreversible oxidation step, and the intermediacy of $(PcFe)_2(O_2)$ and the oxenic species PcFeO is necessary to account for the observed kinetic law. PcFeO is a hypothetical, very reactive, Fe(IV) species, which finds a counterpart only in the parallel porphyrin analog, (P)FeO [69].

The reaction rate law has different formulations, depending on the initial concentration of PcFe. In its simplified form, it can be expressed by the equation

$$- \frac{d[PcFe]}{dt} = k'[PcFe]^2 + k''[PcFe][(PcFe)_2O]$$

Both k' and k" are linear functions of the dioxygen concentration. When the dioxygen concentration is brought to zero, after preliminary oxygenation, the spectrum, according to the mathematical treatment of the rate law equation [59], slowly reverts to that of PcFe or $PcFe(S)_2$ (S = solvent or other donor species). The apparent inconsistency between the formulation of the oxygen containing species in DMSO [46] and in H_2SO_4 [42] observed in the literature [60] is now overcome [59].

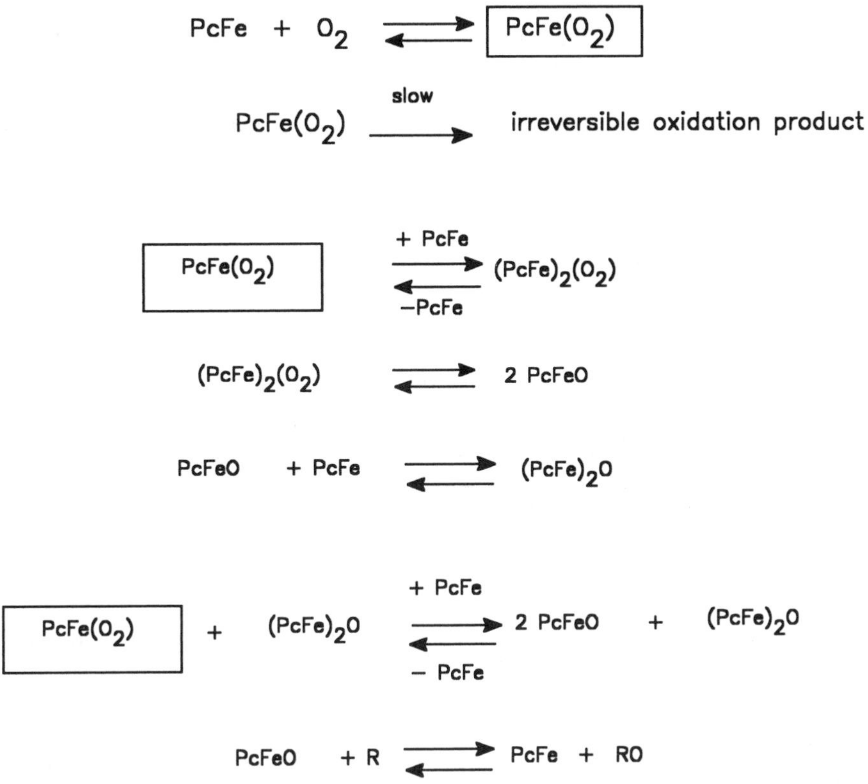

$$PcFe \ + \ O_2 \ \rightleftharpoons \ \boxed{PcFe(O_2)}$$

$$PcFe(O_2) \ \xrightarrow{\text{slow}} \ \text{irreversible oxidation product}$$

$$\boxed{PcFe(O_2)} \ \underset{-PcFe}{\overset{+ \ PcFe}{\rightleftharpoons}} \ (PcFe)_2(O_2)$$

$$(PcFe)_2(O_2) \ \rightleftharpoons \ 2 \ PcFeO$$

$$PcFeO \ + \ PcFe \ \rightleftharpoons \ (PcFe)_2O$$

$$\boxed{PcFe(O_2)} \ + \ (PcFe)_2O \ \underset{- \ PcFe}{\overset{+ \ PcFe}{\rightleftharpoons}} \ 2 \ PcFeO \ + \ (PcFe)_2O$$

$$PcFeO \ + \ R \ \rightleftharpoons \ PcFe \ + \ RO$$

Scheme IV

Reactivity. Both μ-Oxo(1) and μ-Oxo(2) react with pyridine and with other similar N-bases [28, 70]. Prolonged contact with pyridine leads to the breaking of the Fe-O-Fe moiety and formation of the Fe(II) bispyridine adduct, $PcFe(py)_2$. Under controlled contact of either μ-Oxo(1) or μ-Oxo(2) with pyridine, the intermediate adduct of formula [(py)PcFe-O- FePc(py)] can be isolated. From this, the elimination of pyridine gives systematically and reproducibly μ-Oxo(2) [28, 70]. The reaction with pyridine can be summarized as follows:

$$\begin{array}{c} \mu\text{-Oxo}(1) \\ \\ \mu\text{-Oxo}(2) \end{array} \xrightarrow[\text{20h}]{\text{py}} [(\text{py})\text{PcFe-O-FePc(py)}] \xrightarrow{\quad} \begin{array}{l} \xrightarrow{\text{py}} \text{PcFe(py)}_2 \\ \\ \xrightarrow{-\text{py}} \mu\text{-Oxo}(2) \end{array}$$

a. Stoichiometric oxygen atom transfer. $(\text{PcFe})_2\text{O}$, suspended in pyridine/toluene under N_2, transfers quantitatively the oxygen atom to triphenylphosphine as follows [71]:

$$(\text{PcFe})_2\text{O} + \text{PPh}_3 \xrightarrow{\text{py}} 2 \ \text{PcFe(py)}_2 + \text{OPPh}_3 \qquad (1)$$

Using $(\text{PcFe})_2{}^{18}\text{O}$, ${}^{18}\text{OPPh}_3$ has been isolated. In summary, then, dioxygen uptake and oxygen atom transfer can be obtained, under mild conditions (room temperature and 1 atm oxygen pressure), as follows:

$$2 \ \text{PcFe} \xrightarrow{\text{O}_2} (\text{PcFe})_2\text{O} \xrightarrow[\text{PPh}_3]{\text{py}} 2 \ \text{PcFe(py)}_2 + \text{OPPh}_3 \qquad (2)$$

In the absence of added triphenylphosphine, it has been shown that oxygen atom transfer does not take place internally to the phthalocyanine molecule (reaction 1). It is likely that pyridine is the oxygen atom acceptor. The identification of the oxidized species was prevented by practical difficulties.

b. Catalytic oxygen atom transfer. The oxygen atom transfer process from $(\text{PcFe})_2\text{O}$ to triphenylphosphine in toluene and in the presence of pyridine is catalytic at room temperature, when the dioxygen pressure reaches 60-70 atm [50]. A catalytic cycle has been devised from PcFe, as illustrated in Scheme V.

It has been shown that the species PcFe(N-base)_2 are also effective as oxygen atom transfer catalysts. Results obtained with PcFe and a number of bis adducts are reported in Table 4.

PcMn and PcCo derivatives are catalytically inactive. Catalysis is heterogeneous with PcFe and PcFe(im)_2 – which are insoluble in toluene – homogeneous with all the other compounds. No effect of radical inhibitors has been observed. A μ-oxo dimer, formulated as HPcFe-O-FePcH, likely to be $(\text{PcFe})_2\text{O}$, has been found to have low catalytic activity in the oxidation of oxalic acid [72].

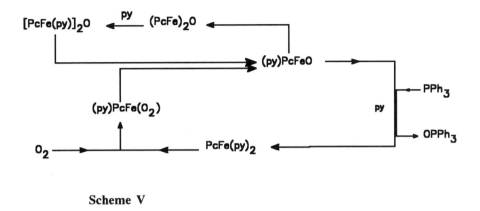

Scheme V

Table 4 Catalytic Activity of PcM (M = Fe, Co, Mn) and of Some Related Species for Oxidation of PPh₃[a]

Catalyst	Molar %[b]	Reaction time (h)	Turnover number[c]
PcFe	1.76	22	14-20
	2.00	22	24
μ-Oxo(1) (4a)	1.74	22	3
μ-¹⁸Oxo(1)	1.74	22	3
PcCo	1.76	22, 44	0
PcMn	1.76	22	0
PcFe(py)₂	0.12	22	125-270
PcFe(im)₂	_[d]	22	11-18
PcFe(1-meim)₂	0.50	22	60-131
PcFe(4-mepy)₂	0.73	22	48-69
PcFe(pip)₂	0.51	22	88-128

[a] Reaction conditions, T = 20-25°C, $p(O_2)$ = 60-70 atm.
[b] With respect to PPh₃.
[c] Molar ratio OPPh₃/catalyst.
[d] Insoluble in the reaction medium.

Electrochemical behavior. The electrochemical oxidation of (PcFe)₂O in pyridine, where indeed the species present is (py)PcFe-O-FePc(py), is consistent with a quasi-reversible metal centered abstraction of one electron, occurring at +0.47 V (*versus* SCE), with the formation of the cationic species [(PcFe)$_2$O]$^+$ [73]. A second electroxidation process at +0.87 V has not been completely analyzed because of its proximity to the edge of the solvent potential window. Sweeping the potential negatively, three electroreduction processes could be observed at −0.59 V, −0.95 V, and the third one at more negative potentials. A parallel electrochemical investigation has also been carried out on the TPP analog, [(TPP)Fe]$_2$O. Midpoint potentials for both μ-oxo dimers are given in Table 5. The only possible comparison between (PcFe)$_2$O and [(TPP)Fe]$_2$O concerns the first monoelectronic reduction process, which indicates that the Pc complex is much more easily reducible than the TPP analog. On the other hand, (PcFe)$_2$O is oxidized to the +1 charged species only slightly less easily than the corresponding Mn compound.

Table 5. Midpoint Potentials (V) for μ-X Dimers in Pyridine

Dimer	Redox Reaction[a]					Ref.
	2+/1+	1+/0	0/1−	1−/2−	2−/3−	
(PcMn)$_2$O (**3**)		0.35	−0.85[b]			[27]
(PcFe)$_2$O (**4**)	0.87	0.47	−0.59	−0.95		[73]
[(TPP)Fe]$_2$O			−1.03	[b]		[73]
(PcFe)$_2$N (**5**)		0.00	−0.83	−1.02	−1.29	[76]
[(TPP)Fe]$_2$N		−0.25	−1.04	−1.52	−1.69	[76]
[(TPP)Fe]$_2$C	1.03	0.52	−1.52[b]			[86]

[a] Denotes charge on the dimer for the electrode reactant and product.
[b] Multielectron process.

PcFe has been extensively studied in electrocatalytic processes, particularly in connection with its activity for O$_2$ reduction in aqueous media (see refs. in [74]). Isolation and characterization of μ-Oxo(1) and μ-Oxo(2) have allowed a more complete understanding of the materials formed when PcFe is deposited from solutions of different solvents on high-surface carbon supports [74]. Fourier-transform IR (FTIR), Mössbauer spectroscopic, and cyclic voltammetric experiments have been performed by using the supported materials and measuring their activity for O$_2$ reduction. Mössbauer spectra indicate that samples of PcFe deposited on carbon-type Vulcan XC-72, prepared from solutions in DMF or DMSO, contain μ-Oxo(1), whereas carbon dispersions obtained from sulfuric acid solutions very likely contain μ-Oxo(2). It is observed that previously studied materials formed by supporting PcFe on Vulcan XC-72 from DMF, DMSO, or pyridine were responsible for doublets believed to be due to PcFe molecules interacting with the carbon substrate. On the basis of the isomer shift and quadrupole splitting values observed, it is suggested that they might be associated with the presence of μ-Oxo(2). Contact of the supported PcFe with O_2 enriches the specimens in the μ-oxo species. Thin porous coating Teflon- bonded electrodes containing small crystals of either monomeric PcFe or any of the two μ-oxo-PcFe species, mixed with high- area carbon, give a common set of well-defined voltammetric peaks in alkaline media. The potentials associated with these peaks are $E_1 = -0.15$ V and $E_2 = -0.55$ V (*versus* Hg/HgO, OH⁻). Rotating ring-disk measurements for O_2 reduction experiments yielded very similar results for both μ-oxo species and PcFe. In the latter case, however, the electrocatalytic and redox features of all forms of PcFe examined are associated with a single PcFe surface species. No spectroscopic evidence has been reported for the presence of O_2 bound to PcFe at room temperature. The strength of the axial interactions between the Fe(II) center and O_2 and its reduction intermediates and products is expected to depend quite critically on the nature of the orbitals involved.

v. $(PcFe)_2N$ and $[(PcFe)_2N]^+$

$(PcFe)_2N$ (**5**) is prepared in 1-chloronaphthalene from PcFe and NaN_3, probably through decomposition of an unstable intermediate azide [75, 76]. Its preparation has also been performed by thermal treatment of $(PNP)[PcFe(N_3)_2]$ [77]. **5** must be formally indicated as a bimetallic mixed-valence Fe(III)-Fe(IV) system with a -3 charge residing on the bridging N atom, PcFe(III)-N-Fe(IV)Pc. Most likely, π-electron conjugation arranges the Fe-N-Fe bond system linearly in **5**, as proved for the only porphyrin analog known by x-rays, that is $(TPPFe)_2N$ [78]. Noticeably, ν_{as}(Fe-N-Fe) in the IR spectra of **5** and $(TPPFe)_2N$ are found very close to one another (915 cm⁻¹

Figure 8 X-ray powder spectra of μ-Oxo(2) (above) and $(PcFe)_2N$ (below).

[76] and 910 cm^{-1} [79], respectively). $(PcFe)_2N$ is isomorphous with μ-Oxo(2) (**4b**) [75], as shown by the x-ray powder spectra given in Figure 8. This reinforces the suggested linearity of Fe-O-Fe in **4b** [as well as in $(PcMn)_2O$)] [28]. The Mössbauer spectrum of **5** [77, 80] shows only a clean single doublet, with data (δ =0.06 mm s^{-1}, ΔE_Q = 1.76-1.78 mm s^{-1}, Table 3) indicative of undifferentiated Fe centers having intermediate oxidation states, between 3 and 4 [i.e., $Fe(III_{1/2})$]. $(PcFe)_2N$ is the first reported derivative of PcFe in which a formal oxidation state higher than +3 has been found. The room temperature magnetic moment of this dimer (μ_B = 2.13) has been reported [80] as indicative of the presence of one unpaired electron per dimer unit, suggesting strong spin coupling. These data, coupled with the EPR spectrum showing a typically axially symmetric structure with $g_{//}$ = 2.03 and $g_{\perp}$ = 2.13, have led to the assignment of a low-spin ground state. A room temperature slightly higher μ_B value (2.33), strongly temperature dependent (1.36 μ_B at 4.2 K), has suggested intramolecular, or intermolecular, spin coupling [77].

$(PcFe)_2N$, like $(PcFe)_2O$ (**4**), is highly thermally stable. In contrast to **4**, $(PcFe)_2N$ does not form solid stable N-base adducts, nor is it converted, in the presence of moderately strong N-bases, like pyridine, into the Fe(II) monomeric species, $PcFe(N\text{-}base)_2$ [76]. Axial coordination of anionic ligands to $(PcFe)_2N$ to form $(TBA)[\{(X)PcFe\}_2N]$ (X = CN, N_3, OCN, SCN, F, OH) has been mentioned [77, 81].

Cyclic voltammetric studies [76], carried out in pyridine solution, show a reversible metal centered one-electron oxidation, at 0.00 V (*versus* SCE), with formation of the cationic species $[(PcFe)_2N]^+$ (**7**). Within the accessible range of the solvent, three-electron reduction transfer processes have also been observed, assumed to occur at the metal centers. The first reduction process is a reversible one-electron transfer. The following two processes maintain their reversibility only at high variable potential sweep rates; low rates determine partial breaking off of the dimer into monomeric Fe(II) species with the final formation of $PcFe(py)_2$. Scheme VI summarizes the reversible processes undergone by the dimer. Parallel observations were conducted on the TPP μ-nitrido analog, with similar results.

$$[(PcFe^{IV,IV})_2N]^+ \quad + \; e \quad \underset{}{\overset{0.00\text{ V}}{\rightleftharpoons}} \quad (PcFe^{III1/2,III1/2})_2N$$

$$(PcFe^{III1/2,III1/2})_2N \quad + \; e \quad \overset{-0.83\text{ V}}{\rightleftharpoons} \quad [(PcFe^{III,III})_2N]^-$$

$$[(PcFe^{III,III})_2N]^- \quad + \; e \quad \overset{-1.02\text{ V}}{\rightleftharpoons} \quad [(PcFe^{II1/2,II1/2})_2N]^{2-}$$

$$[(PcFe^{II1/2,II1/2})_2N]^{2-} \quad + \; e \quad \overset{-1.29\text{ V}}{\rightleftharpoons} \quad [(PcFe^{II,II})_2N]^{3-}$$

Scheme VI

The midpoint potentials, given in Table 5, show systematically higher values of the redox potentials for (PcFe)2N with respect to those found for [(TPP)Fe]2N, indicating a general enhanced tendency towards oxidation for TPP complexes in comparison with Pc complexes of identical charge.

Monoelectronic oxidation of **5**, chemically produced using tetra-cyanoethylene or the ferricinium cation, gives the species [(PcFe)2N](PF6).(H2O)2.(Me2CO), containing the cation [(PcFe)2N]$^+$ [80]. Effective coordination of water and acetone at the Fe centers is uncertain in this species. Its Mössbauer spectrum shows a clean doublet, implying the presence of only one type of Fe. Isomer shift and quadrupole splitting values (Table 3) are very close to those observed for the μ-nitrido TPP analog, [(TPPFe)2N](ClO4) [82]. The negative δ value (-0.10 mm s^{-1}) definitely proves the presence of Fe(IV) in the complex. Accordingly, the IR spectrum excludes the presence of the phthalocyanine radical species Pc(-1)˙ [80].

Briefly reported is the complex [(PcFe)2N]I3 (**7b**). It shows a Mössbauer spectrum with one doublet having $\delta = -0.11$ mm s^{-1} and $\Delta E_Q = 1.98$ mm s^{-1}, definitely indicative of the presence of Fe(IV) [83].

vi. (PcFe)2C

(PcFe)2C (**6**) is obtained in 1-chloronaphthalene by reacting PcFe with CI4 in the presence of sodium dithionite [84, 85], with a procedure very similar to that used for the only μ-carbido dimer known in the field of porphyrins or other porphyrin-like systems. [i.e., (TPPFe)2C] [7]. The IR spectrum of **6** shows an intense absorption at *ca.* 990 cm^{-1} (Table 2), assigned as ν_{as} of the Fe-C-Fe bond system. This absorption is at higher frequency with respect to that found for (TPPFe)2C (940 cm^{-1} with a shoulder at 883 cm^{-1}, [7]), demonstrating a higher degree of π conjugation in **6**. When the frequency is compared with those found for the corresponding μ-oxo (**4a**) and μ-nitrido (**5**) complexes (852, 824, and 910 cm^{-1}, respectively), the π-bond character is in the expected order **6 > 5 > 4**. Stable in air and in a N2 atmosphere up to temperatures of 280-300°C, **6**, like **5** and **4**, is insoluble in most solvents; it dissolves sparingly in N-bases, forming, like **4**, solid stable N- base adducts. The Mössbauer spectrum of **6** has been examined concomitantly by different authors [83, 85] with similar results (Table 3). The negative value observed for the isomer shift clearly indicates the presence of Fe(IV) in the dimer, which forces an assignment of a -4 charge to the bridging C atom, reluctantly accepted [85] and in need of further Mössbauer studies [83]. Significantly high and not adequately explained is the observed value of the quadrupole splitting, the highest observed among the Fe(IV) derivatives of N4 macrocyclic systems. A number of derivatives with anionic ligands confirm the tendency of the Fe centers in **6** to reach six-coordination upon ligation along the axial external free site, either with Pc(-2) or with the radical cation Pc(-1)˙ [83].

vii. (PcRu)₂N

[(PcRu)₂N] **(8)**, [Ru(III₁/₂)], is the only existing example so far reported of a Ru-phthalocyanine complex containing the metal in an oxidation state higher than II. Furthemore, it represents the first N-atom bridged dimeric ruthenium complex carrying macrocyclic porphyrin-like ligands. It is prepared by heating a suspension of accurately purified PcRu [or PcRu(py)₂] in 1-chloronaphthalene in the presence of sodium azide [87]. **8**, stable to air and highly thermally stable (Table 2), insoluble in a variety of solvents, shows no tendency to coordinate N-base molecules at the axial sites of the ruthenium atoms [87]. Its nature as a single-atom bridged dimer has been established mainly on the basis of IR spectral data and x-ray information. X-ray powder patterns have established isomorphism of **8** with the analogous μ-nitrido and μ-oxo Fe species **5** and **4b**. This suggests linearity of the Ru-N-Ru fragment. The assignment of the absorption at 1040 cm⁻¹, found in the IR spectrum of **8** (Table 2) as ν_{as}(Ru-N-Ru) is confirmed by ¹⁵N isotopic enrichment with NaN₃ (99% in ¹⁵N). The labeled (PcRu)₂¹⁵N exhibits this band at 1023 cm⁻¹ (Figure 9). This vibration falls in the range 1150-1000 cm⁻¹, as typically found for linear Ru-N-Ru bond systems [88]. Visible solution and nujol mull IR spectra exclude the presence in **8** of the π-radical ligand Pc(−1)·. Thus, the complex

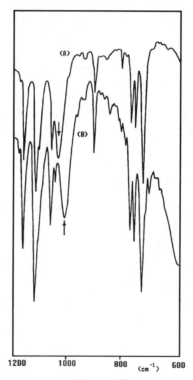

Figure 9 IR spectrum of (PcRu)₂N. (A) ¹⁴N, (B) ¹⁵N. (Reproduced from [87], with permission.)

is formulated as a mixed valence Ru(III)-Ru(IV) dimer. μ_B at room temperature is 1.8, indicating the presence of one unpaired electron per dimer [87], as might be expected for a low-spin complex.

C. SIX-COORDINATE DIMERS

   i. *X-ray work*. It seems appropriate to discuss first the molecular features of the six-coordinate μ-X dimers for which the structure has been elucidated by single crystal x-ray work, that is, the two μ-oxo species [(py)PcMn(III)]$_2$O (**13**) and [(1-meim)PcFe(III)]$_2$O (**20**), and the μ-carbido and μ-nitrido dimers [(1-meim)PcFe(IV)]$_2$C (**29**) and [(Br)PcFe(IV)]$_2$N (**22**) [24, 59, 84, 89]. In view of the difficulty normally found in the preparation of single crystals suitable for x-ray investigation, these structures provide a good deal of information in the field of the dimeric systems examined here. This is of particular importance

Table 6 Crystal Data of Metal Phthalocyanine Dimers and Related Molecules [bond distances (Å) and angles (deg)]

Complex[a]	M-X	M-N$_e$	M-Y$_a$	M-C$_t$	M-X-M	Refs.
[(py)PcMn]$_2$O (**13**)	1.71	1.97	2.15	0.02	178	[24]
[(1-meim)PcFe]$_2$O (**20**) (in I)	1.749	1.92	2.039	0.02	175.1	[59]
[(1-meim)PcFe]$_2$C (**29**) (in II)	1.69	1.95	2.102	0.0	178	[59, 84]
[(Br)PcFe]$_2$N (**22**)	1.639	1.945	2.495	0.0	180	[89]
[(TPP)Fe]$_2$O	1.763	2.087		0.50	174.5	[78]
[(C$_{22}$H$_{22}$N$_4$)Fe]$_2$O	1.792	2.054		0.698	142.7	[98]
[(salen)Fe]$_2$O	1.82	2.10		0.56	139	[99]
[PcFe(1-meim)$_2$] (in I)		1.940	1.946	0.0		[59]
[PcFe(1-meim)$_2$] (in II)		1.932	2.012	0.0		[59, 84]
[PcFe(4-mepy)$_2$]	1.934	2.040		0.0		[100, 101]

[a] X = central bridging atom; N$_e$ = equatorial N$_{(Pc)}$; Y$_a$ = external axial ligand; C$_t$ = center of the N$_4$ chromophore.

when it is considered that six-coordinated analogs are not known in the field of porphyrin or porphyrin-like systems. Crystal data are summarized in Table 6, with additional data on related species. Figures 10-13 illustrate some of the aspects of the molecular arrangement in the structures of **13**, **20**, **22**, **29**.

The structure of **13** was the first one solved [24] and has been, for more than two decades, the only one available on six-coordinate μ-X dimers and, hence, is often referred to in the literature, particularly in connection with Fe(III), mainly five-coordinate, μ-oxo dimers (see, for instance, [90-93]). It seems now to be well established [26] that **13** is the final product isolated from the reaction of PcMn with O_2 in pyridine solution (Scheme I). Loosely held additional molecules of pyridine are normally occluded in the crystal lattice of this complex during preparation [23, 24] and can be easily eliminated by mild heating.

Some details of the structure of the Mn dimer **13** are given in Figure 10. The structure shows two (PcMn) units held together by a bridging oxygen

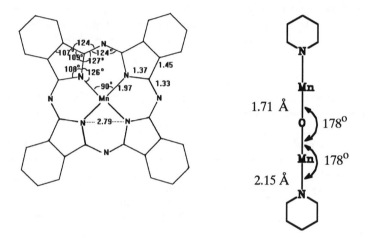

Figure 10 Structural features of [(py)PcMn]2O, (**13**). (Reproduced from [24], with permission.)

atom. The Mn-O-Mn bond system is practically linear (178°) with Mn-O bond distances of 1.71 Å, shorter than expected for a single covalent Mn-O bond. The two phthalocyanine units, relatively rotated by 49° in order to minimize steric and electronic repulsions, are almost parallel to one another with some deviation from planarity of the peripheral benzene rings. Both Mn atoms within the molecule are six coordinated, with the pyridine molecules ligated axially and externally in the sixth coordination position with a Mn-N(py) distance of 2.15 Å.

Single crystal x-ray studies have been carried out on the the two isomorphous species {[(1-meim)PcFe(III)]2O}.[PcFe(II)(1-meim)2].3Me2CO (I) [59] and {[(1-meim)PcFe(IV)]2C}. [PcFe(II)(1-meim)2].3Me2CO (II) [84] formed, respectively, by cocrystallization of the μ-oxo dimer **20** and the μ-

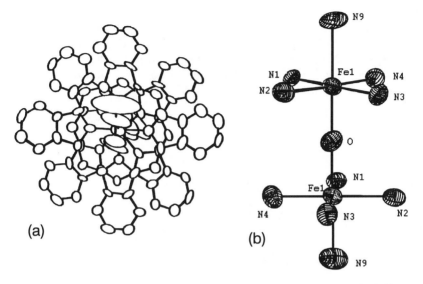

Figure 11 Structure of [1-meim)]PcFe]₂O (in I) as: (a) viewed down the Fe-O-Fe axis: (b) viewed approximately normal to the Fe-O-Fe bond axis. (Reproduced from [59], with permission.)

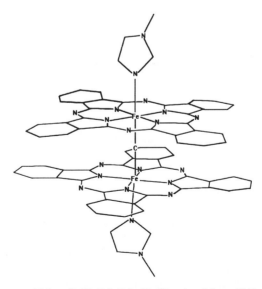

Figure 12 Structure of [(1-meim)PcFe]₂C in II. (Reprinted from [84], with permission.)

carbido dimer **29**, with the monomeric Fe(II) bis adduct PcFe(1-meim)$_2$. Figures 11 and 12 show some details of the dimers present in **I** and **II**. Common relevant features of the structures are: (a) in-plane position of the Fe centers and quasi linear Fe-X-Fe bond moiety; (b) coordination of 1-meim at the external axial sites of the two Fe atoms of the dimer, which are therefore six-coordinated; (c) relative rotation (*ca.* 45°) of the two phthalocyanine molecules within the dimer; (d) eclipsed arrangement of the plane of N-meim with respect to the N$_{(Pc)}$-Fe-N$_{(Pc)}$ equatorial axis. In many respects, the structures show strong analogies with that of the Mn μ-oxo dimer [24]. Similar features are also presented by the bromide complex **22** (Figure 13) [89], apart from point

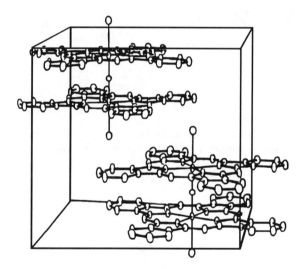

Figure 13 Stacking of the molecules of [(Br)PcFe]$_2$N (**22**). (Reproduced from [89], with permission.)

(d), which is not pertinent. Different features are, instead, presented by the three μ-oxo dimers given in Table 6 and containing five-coordinate Fe(III), representative of the large class of μ-oxo porphyrin or porphyrin-like Fe(III) dimers described in the literature. It has been assumed [85] that points (a)-(c) apply to all the Fe N-base adducts listed in Table 1 (**18-21, 23-30**), as it is reasonably also true for the Mn analogs (**13-17**). Further x-ray information, which reinforces the assumption made, consists of the isomorphism found between the Mn and Fe py adducts **13** and **18**; of importance, due to the known structure of the former. The two corresponding 4-mepy derivatives **16** and **19** are also isomorphous, suggesting a structure very similar to **13** and **18**, since substitution of py by 4-mepy pyridine should not change significantly the general features of the two dimers. As to point (d), established in **13**, **20**, and **29**, it might apply for all the N-base derivatives, with some uncertainty for the pip adducts, owing to the non planar arrangement of this N-base. It seems reasonable to extend points (a), (b), and (c) to all the remaining six-coordinate species reported in Table 1, thus assuming for all of them, as strongly structurally supported, the single-atom bridged molecular framework.

The very close Fe-N$_{(Pc)}$ bond distances (Table 6) observed for the μ-oxo (1.92 Å) and μ-carbido (1.95 Å) Fe dimers and the low-spin (d^6) PcFe(1-meim)$_2$ present in **I** and **II** (1.940, 1.932 Å), and for PcFe(4-mepy)$_2$ (1.932 Å), seem to indicate a low-spin state for the μ-X dimers [59], as confirmed by other data (see the following) [85]. An Fe-C bond distance of 1.69 Å in the μ-carbido bridged dimer, shorter than the Fe-O distance in the μ-oxo analog (1.75 Å), indicates a higher degree of π bonding in the former, in line with expectation. Accordingly, the Fe-N$_{(base)}$ bond distance is longer for the μ-carbido complex than for the μ-oxo analog, due to the *trans* effect. Noticeably, the Fe-N-Fe bridging system in the bromide complex **22** shows the shortest Fe-X bond distance observed (1.639 Å), thus determining a short interplanar distance within the dimer (3.278 Å); a fact that has to be related to the high value of the electrical conductivity measured for this compound ($\sigma_{RT} = 0.5$ ohm^{-1} cm^{-1}) [89].

ii. *Synthesis and properties*. The pyridine adduct **13** is easily obtained by aerial oxidation of a solution of PcMn in the N-base, and subsequent complete evaporation of the latter [21, 23, 25, 26]. Complexes having identical composition (**14-17**) have been prepared by similar procedures (N-base = pip, 3-mepy, 4-mepy, and 1-meim) [25, 70]. The piperidine adduct contains two additional molecules of water and corresponds to the formula [(pip)PcMn]$_2$O.2H$_2$O [25].

The room temperature magnetic moment of **13** is variable from sample to sample [22, 23] and generally lower than the value expected for one unpaired electron per metal atom of the dimer. Similar low values are also observed for the other N-base adducts mentioned [25]. Much higher values should be found for high-spin Mn(III). It is uncertain whether the low μ_B values observed are due to antiferromagnetic interaction between the two high-spin metal centers within the dimer or a low-spin state is present in these complexes with the low paramagnetism measured due to trace amounts of impurities.

Coordination of N-bases such as py, 4-mepy, pip, and 1-meim at the external axial sites of (PcFe)$_2$O produces a new series of air stable adducts of formula [(N-base)PcFe(III)]$_2$O (**18-21**) [28, 70]. Parallel series of μ-nitrido and μ-carbido adducts with the same N-bases, of respective formulae [(N-base)PcFe(IV)]$_2$N(PF$_6$) (**23-26**) [75, 76, 80] and [(N-base)PcFe(IV)]$_2$C (**27-30**) [85], have been also reported. All the μ-oxo adducts were obtained by controlled contact of solid (PcFe)$_2$O with the vapors of the specific N-base or by dissolution of the same μ-oxo dimer in the liquid N-base and subsequent precipitation of the adduct by addition of nonsolvents. Contact of [{PcFe(IV)}$_2$N](PF$_6$) and (PcFe)$_2$C with liquid N-bases leads to the formation of the other two series. The μ-oxo adducts, particularly those containing py or 1-meim, are often contaminated by the presence of the monomeric Fe(II) adducts, PcFe(N-base)$_2$, quantitatively formed by prolonged exposure to the action of the N-base. The magnetic susceptibility and Mössbauer spectral behavior of the complexes [(N-base)PcFe]$_2$O differ from that exhibited by μ-Oxo(1), μ-Oxo(2), and the large number of five- and six-coordinated Fe(III) μ-oxo dimers reported in the literature and indicate the presence of low-spin, weakly antiferromagnetically interacting Fe(III) pairs [70]. The magnetic data are summarized in Table 7. The J values are much lower than those found for μ-Oxo(1) (120 cm^{-1}), μ-Oxo(2) (195 cm^{-1}), and the other Fe(III) high-spin μ-oxo dimers, with insignificant variations within the series. Instead, the magnetic behavior is reminiscent of that observed for the monomeric Fe(III) low-spin [PcFe(OH)$_2$]$^-$ [66].

Table 7 Magnetic Properties for [(N-base)PcFe]$_2$O Complexes[a]

N-base	μ_B[b] (295 K)	g (± 0.1)	J (cm^{-1}) (± 0.2)
4-mepy (**19**)	1.99	2.2	-5.9
pip (**21**)	2.11	2.4	-6.3
1-meim (**20**)	2.16	2.35	-5.5
py (**18**)	1.86	2.16	-5.5

[a] [70].
[b] Magnetic moment per Fe atom.

Isomer shift and quadrupole splitting values at 4.2 K for the adducts **18-21** are within the ranges 0.17-0.20 and 1.58-1.76 mm s^{-1}, respectively, different from those of μ-Oxo(1) (δ = 0.37 mm s^{-1}, ΔE_Q= 0.44 mm s^{-1}) and μ-Oxo(2) (δ = 0.25 mm s^{-1}, ΔE_Q = 1.27 mm s^{-1}) (Table 3, Figure 14). These differences are ascribed to the axial perturbation introduced by the coordinated N-bases and the formation of the six-coordinate environment. It is observed [70] that there is a regular trend in going from μ-Oxo(1) [five-coordinate, bent Fe-O-Fe, high-spin Fe(III)], through μ-Oxo(2) [five-coordinate, linear or quasi-linear Fe-O-Fe, high-spin Fe(III)], to the N-base adducts [six-coordinate, linear Fe-O-Fe, low-spin Fe(III)]. The Mössbauer data of the N-base adducts

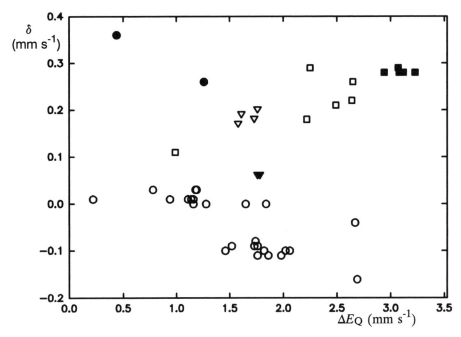

Figure 14 Summary of the Mössbauer data of the μ-X dimeric (a) and some selected Fe(III) monomeric (b) PcFe derivatives. (a) ● , High-spin Fe(III) dimers [μ-Oxo(1) and μ-Oxo(2)]; ▽, low-spin Fe(III) dimers; ▼, Fe(III$_{1/2}$) [(PcFe)$_2$N]; O , Fe(IV) (b) ▢ , Six-coordinate low-spin Fe(III) [66]; ■ , intermediate-spin or spin-admixed Fe(III) (3/2, 5/2) [63, 66].

are similar to those of Fe(III) bis-hydroxy anion [PcFe(OH)$_2$]$^-$, also six-coordinate and low-spin [66].

Applied field Mössbauer spectra have been reported at 4.2 K on the low-spin complex [(4-mepy)PcFe]$_2$O (**19**) [94]. By increasing the field strength from zero to 32.3 kOe, the symmetric doublet observed progressively splits into a triplet (low velocity) and a doublet (high velocity). Hyperfine splitting is not observed, and this is taken as indicative of antiferromagnetic exchange, leading to the coupling of the two $S = 1/2$ Fe(III) centers and to an $S = 0$ ground state. These results are confirmed by the fitting obtained with a simulation of the high-field spectrum if an $S = 0$ ground state is assumed.

Mössbauer spectral data for the μ-nitrido and μ-carbido series of adducts are clearly indicative of the presence of Fe(IV) in both series ($\delta = -0.10$ - 0.00 mm s^{-1} [80, 85], Table 3, Figure 14). The quadrupole splitting values of the μ-nitrido series [80] (1.76 - 1.52 mm s^{-1}) are also in close range, practically coincident with that of the μ-oxo adducts (1.76 - 1.58 mm s^{-1}), with the same order of decreasing values (4-mepy $\approx$ py > pip > 1-meim). This order is also maintained with the series of μ-carbido adducts, which, however, show a significantly lower range of values (1.19 - 0.94 mm s^{-1}). Mössbauer spectra, together with magnetic susceptibility data, NMR, and structural information concur to identify the two series of μ-nitrido and μ-carbido adducts as low-spin Fe(IV) six-coordinate species, as found for the μ-oxo analogs [85].

μ-Nitrido Fe dimers having coordinated anions and containing Pc(-2) or the radical Pc(-1)· have also been prepared and reported [77]. Oxidation of (PcFe)$_2$N with Br$_2$, CF$_3$CO$_2$H , or concentrated HNO$_3$ produces the series [{(X)PcFe(-1)}$_2$N]X (X = Br, **33**; CF$_3$CO$_2$, **34**; NO$_3$, **35**). The Mössbauer spectra of these species show a single quadrupole doublet and, thus, indicate the presence of only one type of Fe center. The negative isomer shift values (δ = -0.08 - -0.10 mm s^{-1}) clearly suggest an oxidation state of $+4$ for Fe. The bromide complex shows a single doublet that can be best fitted to two doublets, suggesting slightly different Fe(IV) sites. The IR spectra exhibit absorptions at 1459 and 1350 cm^{-1}, which indicate the presence of the π-radical ligand Pc(-1)·. Magnetic moments and ESR spectra of the bromide and trifluoroacetate complexes point to an S = 1/2 ground state. Similarly formulated μ-nitrido dimers with X = Cl or RCO$_2$, preliminarly presented [81], have not yet been definitely reported. This is also the case for a number of complexes of formula [{(X)PcFe}$_2$N]$^{2-}$ (X = CN, N$_3$, OCN, SCN, F, OH) [81].

A series of Fe(IV)-Pc(-2) complexes and Fe(IV)-(Pc(-1))· μ-carbido complexes with coordinated anions have been reported [83]. Once again, the presence of Fe(IV) in these complexes is proved by their Mössbauer spectra. In fact, the isomer shift values are all very close to zero, in keeping with expectation. The presence of the Pc(-1)· π-radical ligand is shown by IR and visible spectra and electrochemical properties.

D. SUBSTITUTED ANALOGS

The synthesis of the μ-oxo species (TTBPcFe)$_2$O (**9**), formed by tetra(*tert*-butyl) substituted iron phthalocyanine, as well as its characterization and behavior, have been re-examined in detail [95] in the light of previous literature on the same subject [28, 30, 43, 48, 55-57]. The visible spectral changes observed for (TTBPcFe)$_2$O in pyridine solution follow strictly [95] those observed for the unsubstituted (PcFe)$_2$O [28]. This unequivocally confirms the nature of μ-oxo species of (TTBPcFe)$_2$O, as discussed previously (see Section B).

μ-Oxobis[tetra(dodecylsulfonamido)phthalocyaninatoiron(III)], (**11**) (TDPcFe)$_2$O, and μ-oxobis[tetra(octylsulfonamido)phthalocyaninatoiron(III)], (**10**) (TOPcFe)$_2$O, were obtained by chromatography on neutral alumina of their Fe(II) precursors, using mixtures of 5% methanol in chloroform or 5% methanol in dichloromethane as eluents [30]. These species are identified as

μ-oxo dimers and should be classified, like $(PcFe)_2O$, as high-spin Fe(III) complexes on the basis of spectroscopic and magnetic data. These data include, for $(TDPcFe)_2O$, an IR absorption at 810 cm^{-1}, assigned as ν_{as}(Fe-O-Fe), and a μ_B (RT) value of 2.0, strongly temperature dependent. Similar features are observed for $(TOPcFe)_2O$ [μ_B (RT) = 1.7]. $(TDPcFe)_2O$ in DMF or THF (10^{-5}- 10^{-6} M) is photoreduced to TDPcFe(II), if irradiated, under nitrogen, into its Q band (633 nm). Cleavage of the dimer with formation of the Fe(II) monomer is also determined by the action of electron donors such as imidazole, 2-methylimidazole, and dimethylamine.

E. MIXED DIMERS

Single-atom bridged dimers involving different ligands and/or different metal centers have received recently attention as μ-oxo species, but include only one phthalocyanine derivative, that is (TPP)Cr(III)-O-Fe(III)Pc (**12**) [97].

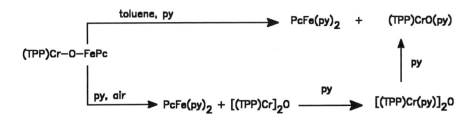

Scheme VII

The synthesis of **12** is performed in 1-chloronaphthalene from TPPCrO and PcFe. It gives equimolar amounts of (TPP)CrCl and PcFeCl by reaction with HCl. Some of the reactivity of **12** is summarized in Scheme VII. It decomposes, in benzene or toluene, to (TPP)CrO and $(PcFe)_2O$ in the presence of O_2. **12** shows a visible absorption spectrum in 1-chloronaphthalene/benzene with maximum intensity absorptions due to the TPP and Pc ligands at 431 nm (log ε = 4.56) and 699 nm (log ε = 4.34), respectively [97]. No IR absorptions attributable to ν_{as}(Fe-O-Fe) are observed in the region 900-800 cm^{-1}. A spin state (5/2, 3/2) is suggested on the basis of its magnetic moment (μ_B = 3.12).

REFERENCES

1. K. S. Murray, *Coord. Chem. Rev.*, **12** (1974) 1.
2. S. J. Lippard, *Angew. Chem., Int. Ed. Engl.*, **27** (1988) 344.
3. B. O. West, *Polyhedron*, **8** (1989) 219.
4. J. P. Collman, T. R. Halpert and K. S. Suslick, in Metal Ion Activation of Dioxygen, T. G. Spiro, Ed., J. Wiley & Sons, Inc., New York, 1980, pp. 1-72.
5. D. F. Bocian, E. W. Findsen, J. A. Hoffmann, jr., G. A. Schick, D. R. English, D. N. Hendrickson, and K. S. Suslick, *Inorg. Chem.*, **23** (1984) 807.
6. D. R. English, D. N. Hendrickson, and K. S. Suslick, *Inorg. Chem.*, **24** (1985) 122.
7. D. Mansuy, J.-P. Lecomte, J.-C. Chottard, and J.-F. Bartoli, *Inorg. Chem.*, **20** (1981) 3119.
8. V. L. Goedken, M. Deakin, and L. A. Bottomley, *J. Chem. Soc., Chem. Commun.*, (1982) 607.
9. N. Jiazan, S. Feng, L. Zhenxiang, and Y.Shaoming, *Inorg. Chim. Acta*, **139** (1987) 165.
10. K. J. Wynne, *Inorg. Chem.*, **24** (1985) 1339.
11. T. Inabe, J. G. Gaudiello, M. K. Maguel, J. W. Lyding, R. L. Burton, W. J. McCarthy, C. R. Kanneworf, and T. J. Marks, *J. Am. Chem. Soc.*, **108** (1986) 7595.
12. M. Y. Ogawa, J. Martinsen, S. M. Palmer, J. L. Stanton, J. Tanaka, R. L. Green, B. M. Hoffman, and J. A. Ibers, *J. Am. Chem. Soc.*, **109** (1987) 1115.
13. C. J. Schramm, R. P. Scaringe, D. R. Stojakovic, B. M. Hoffman, J. A. Ibers, and T. J. Marks, *J. Am. Chem. Soc.*, **102** (1980) 6702.
14. M. Massoyan-Deneux, D. Benlian, M. Pierrot, A. Fournel, and J. P. Sorbier, *Inorg. Chem.*, **24** (1985) 1878.
15. K. Yakushi, M. Sakuda, H. Kuroda, A. Kawamoto, and J. Tanaka, *Chem. Lett.*, (1986) 2261.
16. K. Yakushi, M. Sakuda, I. Hamada, H. Kuroda, A. Kawamoto, J. Tanaka, T. Sugano, and M. Kinoshita, *Synt. Met.*, **19** (1987) 769.
17. M. Almeida, M. G. Kanatsidis, L. M. Tonge, T. J. Marks, H. D. Marcy, W. J. McCarthy, and C. R. Kanneworf, *Solid State Comm.*, **63** (1987) 457.
18. Z. Q. Zeng, R. Aroca, A. M. Hor, and R. O. Loufty, *J. Raman Spectr.*, **20** (1989) 467.
19. (a) D. Guay, G. Veilleux, R. G. Saint-Jacques, R. Côté, and J. P. Dodelet, *J. Mater. Res.*, **4** (1989) 651; (b) D. Guay, J. P. Dodelet, R. Côté, C. H. Langford, and D. Gravel, *J. Electrochem. Soc.*, **136** (1989) 2272.
20. V. L. Goedken, G. Dessy, C. Ercolani, V. Fares, and L. Gastaldi, *Inorg. Chem.*, **24** (1985) 991.
21. J. A. Elvidge and A. B. P. Lever, *Proc. Chem. Soc., London*, (1959) 195.
22. G. Engelsma, A. Yamamoto, E. Markham, and M. Calvin, *J. Phys. Chem*, **66** (1962) 2517.
23. A. Yamamoto, L. K. Phillips, and M. Calvin, *Inorg. Chem.*, **7** (1968) 847.
24. L. H. Vogt, jr., A. Zalkin, and D. H. Templeton, *Inorg. Chem.*, **6** (1967) 1725.
25. G. W. Rayner Canham and A. B. P. Lever, *Inorg. Nucl. Chem. Lett.*, **9** (1973) 513.
26. A. B. P. Lever, J. P. Wilshire, and K. S. Quan, *Inorg. Chem.*, **20** (1981) 761.

27. P. C. Minor and A. B. P. Lever, *Inorg. Chem.*, **22** (1983) 826.
28. C. Ercolani, M. Gardini, F. Monacelli, G. Pennesi, and G. Rossi, *Inorg. Chem.*, **22** (1983) 2584.
29. A. B. P. Lever, J. P. Wilshire, and S. K. Quan, *J. Am. Chem. Soc.*, **101** (1979) 3668.
30. A. B. P. Lever, S. Licoccia and B. S. Ramaswamy, *Inorg. Chim. Acta*, **64** (1982) L87.
31. G. T. Byrne, R. P. Linstead, and A. R. Lowe, *J. Chem. Soc.*, (1934) 1017.
32. R. P. Linstead and A. R. Lowe, *J. Chem. Soc.*, (1934) 1022.
33. C. E. Dent, R. P. Linstead, and A. R. Lowe, *J. Chem. Soc.*, (1934) 1033.
34. J. A. Elvidge and R. P. Linstead, *J. Chem. Soc.*, (1935) 3536.
35. C. E. Dent and R. P. Linstead, *J. Chem. Soc.*, (1934) 1027.
36. P. A. Barrett, C. E. Dent and R. P. Linstead, *J. Chem. Soc.*, (1936) 1719.
37. P. A. Barrett, D. A. Frye, and R. P. Linstead, *J. Chem. Soc.*, (1938) 1157.
38. J. M. Robertson, *J. Chem. Soc.*, (1935) 615.
39. J. M. Robertson, *J. Chem. Soc.*, (1936) 1195.
40. J. M. Robertson and I. Woodward, *J. Chem. Soc.*, (1937) 219.
41. I. Collamati, C. Ercolani, and G. Rossi, *Inorg. Nucl. Chem. Lett.*, **12** (1976) 799.
42. C. Ercolani, F. Monacelli, and G. Rossi, *Inorg. Chem.*, **18** (1979) 712.
43. C. Ercolani, G. Rossi, and G. Monacelli, *Inorg. Chim. Acta*, **44** (1980) L215.
44. C. Ercolani, M. Gardini, K. S. Murray, G. Pennesi, and G. Rossi, *Inorg. Chem.*, **25** (1986) 3972.
45. B. W. Dale, *Trans. Farad. Soc.*, **65** (1969) 331.
46. C. Ercolani, G. Rossi, F. Monacelli, and M. Verzino, *Inorg. Chim. Acta*, **73** (1983) 95.
47. J. G. Jones and M. V. Twigg, *Inorg. Nucl. Chem. Lett.*, **44** (1972) 305.
48. W. Kalz and H. Homborg, *Z. Naturforsch.*, **38b** (1983) 470.
49. C. Ercolani et al., unpublished results.
50. C. Ercolani, M. Gardini, G. Pennesi, and G. Rossi, *J. Mol. Catal.*, **30** (1985) 135.
51. B. J. Kennedy, K. S. Murray, P. R. Zwack, H. Homborg, and W. Kalz, *Inorg. Chem.*, **24** (1985) 3302.
52. I. Collamati, *Inorg. Chim. Acta*, 35 (1979) L303.
53. I. Collamati, *Inorg. Chim. Acta*, 124 (1986) 61.
54. B. Hutchinson, B. Spencer, R. Thompson, and P. Neill, *Spectrochim. Acta*, **43A** (1987) 631.
55. N. I. Bundina, O. L. Kaliya, O. L. Lebedev, E. A. Lukyanets, G. N. Rodionova, and T. M. Ivanova, *Koord. Khim.*, **2** (1976) 940; English Translation, Plenum, (1976) 720.
56. N. G. Mekbryakova, N. I. Bundina, T. Yu. Gulina, O. L. Kaliya, and E. A. Lukyanets, *Zh. Obsh. Khim.*, **54** (1984) 1656.
57. J. Metz, O. Schneider, and M. Hanack, *Inorg. Chem.*, **23** (1984) 1065.
58. C. S. Frampton and J. Silver, *Inorg. Chim. Acta*, **18** (1985) 187.
59. C. Ercolani, F. Monacelli. S. Dzugan, V. L. Goedken, G. Pennesi, and G. Rossi, *J. Chem. Soc., Dalton Trans.*, (1991) 1309
60. I. Collamati, *Inorg. Nucl. Chem. Lett.*, **17** (1981) 69.
61. N. I. Bundina, N. G. Mekhryakova, T. Yu. Gulina, V. V. Zelentsov, A. S. Luppov, O. L. Kaliya, O. L. Lebedev, and E. A. Lukyanets, Summaries of Papers Presented at Sixteenth All Union Conference on the Chemistry of Complex Compounds, Part 1, Ivenovo (1981) 110.

62. J. F. Myers, G. W. R. Canham, and A. B. P. Lever, *Inorg. Chem.*, **14** (1975) 461.
63. B. J. Kennedy, G. Brain, and K. S. Murray, *Inorg. Chim. Acta*, **81** (1984) L29.
64. S. M. Palmer, J. L. Stanton, N. K. Jaggi, B. M. Hoffman, J. A. Ibers, and L. H. Schwartz, *Inorg. Chem.*, **24** (1985) 2040.
65. J. Jones and M. V. Twigg, *J. Chem. Soc. A*, (1970) 1546.
66. B. J. Kennedy, K. S. Murray, P. R. Zwack, H. Homborg, and W. Kalz, *Inorg. Chem.*, **25** (1986) 2539.
67. W. Kalz, H. Homborg, H. Kuppers, B. J. Kennedy, and K. S. Murray, *Z. Naturforsch.*, **39b** (1984) 1478.
68. N. Kobayashi, H. Shirai, and N. Hoyo, *J. Chem. Soc., Dalton Trans.* (1984) 2107.
69. D. H. Chin, G. N. La Mar and A. L. Balch, *J. Am. Chem. Soc.*, **102** (1980) 5945.
70. C. Ercolani, M. Gardini, K. Murray, G. Pennesi, G. Rossi, and P. R. Zwack, *Inorg. Chem.*, **26** (1987) 3539.
71. C. Ercolani, M. Gardini, G. Pennesi, and G. Rossi, *J. Chem. Soc., Chem. Comm.*, (1983) 549.
72. S. A. Borisenkova, Th. Th. Hegele, and V. S. Pozii, *Vestn. Mosk. Univ., Ser. 2: Khim.*, **27** (1986) 70.
73. L. A. Bottomley, C. Ercolani, J.-N. Gorce, G. Pennesi, and G. Rossi, *Inorg. Chem.*, **25** (1986) 2338.
74. A. A. Tanaka, C. Fierro, D. Scherson, and E. B. Yeager, *J. Phys. Chem.*, **91** (1987) 379.
75. V. L. Goedken and C. Ercolani, *J. Chem. Soc., Chem. Commun.*, (1984) 378.
76. L. A. Bottomley, J.-N. Gorce, V. L. Goedken, and C. Ercolani, *Inorg. Chem.*, **24** (1985) 3733.
77. B. J. Kennedy, K. S. Murray, H. Homborg, and W. Kalz, *Inorg. Chim. Acta*, **134** (1987) 19.
78. A. B. Hoffman, D. M. Collins, V. W. Day, E. B. Fleischer, T. S. Srivastava, and J. L. Hoard, *J. Am. Chem. Soc.*, **94** (1072) 3620.
79. D. A. Summerville and I. A. Cohen, *J. Am. Chem. Soc.*, **98** (1976) 1747.
80. C. Ercolani, M. Gardini, G. Pennesi, G. Rossi, and U. Russo, *Inorg. Chem.*, **27** (1988) 422.
81. H. Homborg, B. Peters, K. S. Murray, and E. N. Bakshi, XXVI ICCC, Porto (Portugal), MS-3, P4 (1988).
82. R. English, D. N. Hendrickson, and K. S. Suslick, *Inorg. Chem.*, **22** (1983) 3302.
83. E. N. Bakshi, C. D. Delfs, K. S. Murray, B. Peters, and H. Homborg, *Inorg. Chem.*, **27** (1988) 4318.
84. G. Rossi, V. L. Goedken, and C. Ercolani, *J. Chem. Soc., Chem. Comm.*, (1988) 86.
85. C. Ercolani, M. Gardini, V. L. Goedken, G. Pennesi, G. Rossi, U. Russo, and P. Zanonato, *Inorg. Chem.*, **28** (1989) 3097.
86. D. Lancom and K. M. Kadish, *Inorg. Chem.*, **23** (1984) 3942.
87. G. Rossi, M. Gardini, G. Pennesi, C. Ercolani, and V. L. Goedken, *J. Chem. Soc., Dalton Trans.*, (1989) 193.
88. M. J. Cleare and W. P. Griffith, *J. Chem. Soc. A*, (1970) 1117.
89. B. Moubaraki, D. Benlian, A. Baldy, and M. Pierrot, *Acta Cryst. C*, **45** (1989) 393.
90. S. J. Lippard, H. Shrugav, and C. Walling, *Inorg. Chem.*, **6** (1967) 1825.
91. J. F. Kirner, W. Don, and W. R. Scheidt, *Inorg. Chem.*, **15** (1976) 1685.

92. J. O. Alben, W. H. Fuchsman, C. A. Beaudreau, and W. S. Caughey, *Biochem.*, **7** (1968) 24.
93. J. A. Cohen and W. S. Caughey, *Biochem.*, **7** (1968) 636.
94. E. N. Bakshi and K. S. Murray, *Hyper. Inter.*, **40** (1988) 283.
95. R. Heucher, Dissertation, University of Köln (Germany), 1985.
96. C. Ercolani and R. Heucher, unpublished results.
97. D. J. Liston, B. J. Kennedy, K. S.Murray, and B. O. West, *Inorg. Chem.*, **24** (1985) 156.
98. M. C. Weiss and V. L. Goedken, *Inorg. Chem.*, **18** (1979) 8.
99. M. Gerloch, E. D. McKenkie, and A. D. C. Towl, *J. Chem. Soc. A*, (1969) 2859.
100. T. Kobayashi, F. Furokawa, I. Ashida, N. Uyeda, and E. Suito, *J. Chem. Soc., Chem. Comm.*, (1971) 1631.
101. F. Cariati, F. Morazzoni, and M. Zocchi, *J. Chem. Soc., Dalton Trans.*, (1978) 1078.

2

Mössbauer

Spectroscopy of

Phthalocyaninato-

metal Complexes

M. Hanack, U. Keppeler,
A. Lange, A. Hirsch,
and R. Dieing

A. INTRODUCTION

The first measurements of the recoil free emission and absorption of γ-quants were carried out mainly on inorganic compounds. Already by 1962 measurements on phthalocyaninatoiron(II) (Fig. 1) were published [1].

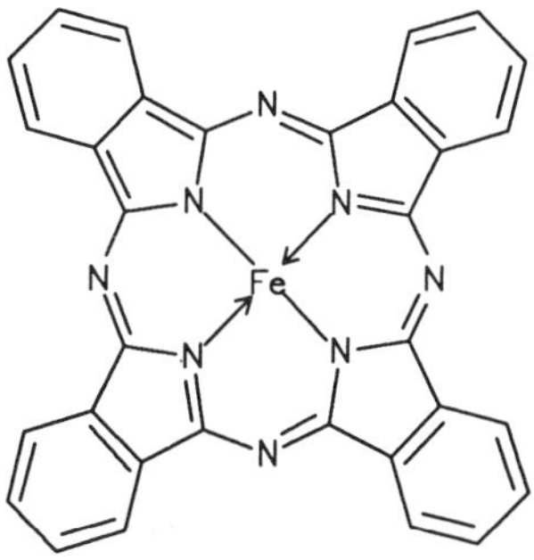

Figure 1 Structure of phthalocyaninatoiron(II).

From quite early on we began to recognize the advantages of this method for our work on bridged macrocyclic transition metal complexes. These materials are of great interest, since the bridged systems (Fig. 2) often show semiconducting properties even without external oxidative doping [2−5]. Some of these compounds [3] also display interesting magnetic and nonlinear optic properties as well as photoconductivity.

Problems arise from the low solubility of many of these complexes in common organic solvents. In addition to ^{13}C solid-state NMR [6], for iron-containing compounds ^{57}Fe Mössbauer spectroscopy is an excellent tool to obtain more information about these solids.

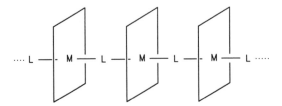

Figure 2 Schematic drawing of bridged macrocyclic transition metal compounds.

Detailed summaries of the Mössbauer spectroscopy of iron porphyrins were presented [7, 8], but only part of the studies published on phthalocyaninatometal complexes were reported. In the following we review Mössbauer spectroscopy of phthalocyaninatometal complexes, including topics such as polymeric bridged phthalocyaninatometal complexes and phthalocyaninatoiron(II) isocyanide complexes, which were not taken into account in the reviews mentioned. Data are presented from its beginning in 1962 up to the present. This chapter has been written from the point of view of an organic chemist who uses Mössbauer spectroscopy to characterize newly synthesized materials and who is interested in their structures and properties; therefore, detailed theoretical interpretations are not given here. For further information concerning the theoretical background of Mössbauer spectroscopy and the instrumentation we refer to the literature [9–16].

Experimental work on the Mössbauer spectroscopy of phthalocyaninatometal complexes may be divided into four sections:

1. Studies on phthalocyaninatoiron complexes.
2. Studies on phthalocyaninatotin complexes.
3. Studies on phthalocyaninatocobalt complexes.
4. Studies on iodine containing phthalocyaninatometal complexes.

As the isotope ^{57}Fe has the most easily observable Mössbauer transition, most of the published papers deal with measurements on phthalocyaninatoiron complexes. The following five subdivisions will be distinguished:

1. Measurements on the square-planar metallomacrocycle PcFe, its substituted derivatives R_xPcFe ($x = 4$, 8, 16; see Table 4), and the corresponding naphthalocyanines NcFe, characterized by an additional condensed benzene ring.
2. Measurements on hexacoordinated PcFe(II)L$_2$ complexes, PcFe(II)LL' complexes, and bridged [PcFe(II)L]$_n$ complexes.
3. Measurements on monomeric and bridged PcFe(III) complexes and hexacoordinated bridged mixed-valence systems.

4.   Dimeric μ-oxo, μ-nitrido and μ-carbido complexes of PcFe.

5.   Investigations directed towards the catalytical activity of monomeric and polymeric PcFe
     complexes in different oxidation states.

Figure 3 shows a simplified schematic representation of the shifts and splittings of the
energy levels of the ^{57}Fe nucleus, resulting from the hyperfine interaction of the monopole,
dipole, and quadrupole moments with the local electric and magnetic field.

The simple spectra of the diamagnetic phthalocyaninatoiron complexes can be described
by the two hyperfine parameters, the isomer shift δ and the quadrupole splitting ΔE_Q. The value
of the isomer shift correlates with the electron density at the nucleus and is affected by the total
s-electron population and the shielding by the $3d$ electrons.

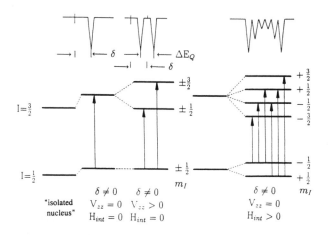

Figure 3 Schematic representation of the splitting and shift of the ^{57}Fe nuclear energy levels.

As a result of a nonspherical charge distribution around the nucleus, the interaction of
the electric field gradient (EFG) with the nuclear quadrupole moment splits the 14.4 keV, I =
3/2 energy level into two sublevels ($m_I = \pm\,3/2, \pm\,1/2$). The EFG at the nucleus may be
described by a symmetric traceless tensor V_{ij}

$$\text{EFG} = -V_{ij} = -\,(\partial^2 V/\partial_i\partial_j), \quad i,j = \text{x, y, z,}$$

where V_{ij} is the electrostatic potential. The axis system of the Mössbauer atom is defined so that
V_{zz} is the maximum value of the field gradient. As a consequence of the Laplace equation,
requiring that the diagonal element of the EFG vanishes ($V_{xx} + V_{yy} + V_{zz} = 0$), the EFG is
described completely by two independent parameters. Generally V_{zz} and the asymmetry parameter
η, defined as

$$\eta = \frac{(V_{xx} - V_{yy})}{V_{zz}}$$

are used. The convention $|V_{zz}| \geq |V_{yy}| \geq |V_{xx}|$ leads to the condition $0 \leq \eta \leq 1$.

By quantum-mechanical calculations the energy of the Mössbauer lines can be calculated, and for ^{57}Fe (ground-state $I = 1/2$, 14.4 keV excited state $I = 3/2$) the spectrum consists of two lines separated by the quadrupole splitting ΔE_Q with

$$\Delta E_Q = \tfrac{1}{2} e \, QV_{zz}(1 + \frac{\eta^2}{3})^{1/2}.$$

In addition to this valence contribution due to the $3d$ valence electrons, two further contributions to the EFG can be distinguished. The lattice part arises from charges on distant atoms; the third component arises from the charge distribution in bonds with covalent character.

If there is a magnetic field at the nucleus, either internal or by an externally applied magnetic field, a third hyperfine interaction, the nuclear Zeeman effect, will cause a further splitting of the lines. For a detailed description of the hyperfine interactions, we refer to the literature [13].

B. STUDIES ON PHTHALOCYANINATOIRON COMPLEXES
i. Measurements on Square-Planar Metallomacrocycles

The first values for the isomer shift δ and the quadrupole splitting ΔE_Q of PcFe(II) were reported in 1962 [1]. In the following period a vast number of measurements were carried out. Various techniques were applied to gain insight into the electronic structure of PcFe. Thus the influence of pressure on the Mössbauer resonance of PcFe has been investigated [17, 18]. The results of measurements on impure PcFe [17] were corrected in a second paper [18]. In addition, the effect of grinding crystalline PcFe and heating the material in the presence of air was described [18]. In a more sophisticated manner the Mössbauer spectra of powdered samples of the intermediate-spin PcFe(II) at low temperature in applied magnetic fields were measured [19]. Further measurements on a single crystal of PcFe confirmed the first results [20]. A positive sign of the electric field gradient at the iron nucleus was established, and a $3d$ electronic ground-state configuration of $(d_{xy})^2 (d_{xz})^1 (d_{yz})^2 (d_{z^2})^1$ was deduced from a crystal-field model. Further attempts to elucidate the electronic structure of PcFe were made via Mössbauer measurements [21] and later via magnetic susceptibility and Mössbauer investigations [22]. As a result of these investigations, an ^3E_g ground-state corresponding to the $(a_{1g})^2 (e_g)^3 (b_{2g})^1$ configuration was proposed. It should be mentioned that commercial-grade PcFe without further purification was used.

In addition to the measurements at low temperatures, the effect of higher temperatures on the Mössbauer spectrum of PcFe was investigated. After treatment of PcFe at 600°C and higher temperatures, the iron-containing pyrolysis products were analyzed [23]. Attempts to prove the existence of an excited state of PcFe were made by measurements up to 557 K, but no additional quadrupole doublet appeared [24]. The Mössbauer spectra of $PcFe^+X^-$ (X = Cl, $p\text{-}CH_3C_6H_4SO_3$), PcFe, and its anionic reduced complexes $Li_n[PcFe]$ ($n = 1-3$) were reported, and a molecular orbital approach was used in order to explain the values for the isomer shifts and the quadrupole splittings as well as their dependence on temperature [25−27]. Many further measurements on PcFe were published. Those dealing with the catalytical activity of PcFe will be discussed in Section B,v. The variation of the isomer shift, the quadrupole splitting, the intensity of the lines, and the linewidth in the temperature range from -160 to +150°C was described for PcFe [28] and for a mixture with the metal-free PcH_2 [29]. The occurrence of different forms of PcFe was reported. They can be converted to the normal PcFe by reaction with pyridine to yield $PcFe(py)_2$ and thermal decomposition of this complex [30, 31].

Using different methods for the preparation of PcFe, two forms having different Mössbauer parameters could be obtained [32]. Although no structural details were given, presumably two different modifications of PcFe(II) have been isolated. The intermolecular interactions in the α and ß modification of PcFe were studied by emission spectroscopy [33] and will be discussed in Section D.

Here we want to add a note concerning the purity of PcFe. Most of the Mössbauer experiments have been carried out with sublimed material, but crude commercial-grade material has also been used. We found out that the purity of the material depends on the synthetic pathway. For the synthesis of monomeric and bridged PcFe complexes, we needed pure starting material. An investigation on a sample of a technical-grade PcFe(II) demonstrated that this material is merely of poor quality. Beside PcH_2 [34] and further organic impurities, in addition Mössbauer spectroscopy proved, that iron-containing impurities are also present [35]. The spectrum was fitted to 3 quadrupole doublets. Beside the doublet of PcFe, the doublet of the $\mu\text{-oxo}(1)$ dimer $(PcFe)_2O$ [36] can be assigned, while the origin of the third, broad doublet is not clear. Probably iron oxides are present.

From our own experience we know [37] that a synthesis from phthalodinitrile and pentacarbonyliron [38] yields a sufficiently pure PcFe. As demonstrated by the Mössbauer spectra, no iron-containing impurities are present (see Fig. 4).

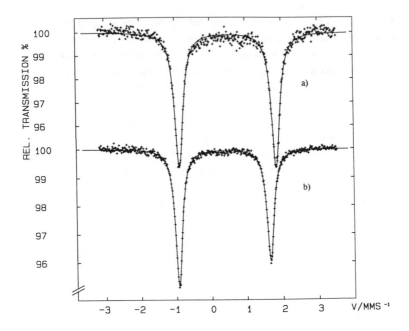

Figure 4 Mössbauer spectra [35] of (a) PcFe (77 K) and (b) sublimed PcFe (293 K; isomer shift relative to metallic iron).

Beside these measurements of the hyperfine parameters of PcFe, investigations of the electronic structure of PcFe provide calculated values of ΔE_Q. As mentioned, attempts were made to explain the measured values of the hyperfine parameters of PcFe, PcFeCl and the anionic complexes Li_n[PcFe] ($n = 1-3$) by a molecular-orbital approach [25]. The calculated quadrupole splitting for PcFe as a sum of a valence contribution of 2.2 mm/s and a lattice contribution of 0.35 mm/s agrees well with the observed value of 2.58 mm/s (298 K). In addition to these considerations, the value of ΔE_Q with respect to the possible ground-state $3d$ electron configurations of PcFe was calculated using a simple crystal-field model [20]. Based on a study of the electron density of PcFe by low-temperature x-ray diffraction an ^3E_gA state [according to a $(d_{xy})^2 (d_{xz},d_{yz})^3 (d_{z^2})^1$ population] was determined as the main contribution to the ground-state and a value of +2.2 mm/s was calculated for the valence contribution and a considerably smaller value of +0.2 mm/s for the lattice contribution [39]. Using semiempirical molecular-orbital linear combination of atomic orbitals (MO-LCAO) calculations, the same ground-state configuration ^3E_gA and very similar d-orbital populations were found [40] though, in this case the tetraazaporhyrinato ligand was used. A quadrupole splitting of 2.89 mm/s was calculated. The principal z-axis for this state is reported to coincide with the C_4 symmetry axis.

A number of phthalocyaninatoiron complexes bearing substituents (R) at the four benzene rings have been synthesized. By means of these 4, 8, or 16 substituents, many properties of the R_xPcFe complexes can be changed, for example, their thermal stability, their oxidation potentials [41], or their solubility. Complexes that are substituted with long chains may show interesting liquid crystalline behavior [42].

The Mössbauer data of the complexes R_xPcFe are listed in Table 1. Most of the spectra are comparable to the unsubstituted PcFe, consisting of one quadrupole doublet, with some variation in the hyperfine parameters, depending on the nature of the substituents. However, some of the complexes are reported to have spectra with more than one quadrupole doublet. These doublets were assigned without further characterization to Fe(III) complexes [43, 44]. Some of the complexes are characterized incompletely so that their Mössbauer data may not be very reliable.

By establishing so-called bonding parameters Σ and Π, calculated from Mössbauer data, attempts were made to correlate the effects of electron-withdrawing [45] and electron-releasing [46] substituents with the Mössbauer parameters. Σ is defined to be larger for a stronger σ donor, while Π is larger for a stronger π donor and for a weaker π acceptor. The Mössbauer data can be related to these relative bonding parameters by two equations [45]:

$$\delta = - \Sigma + \Pi \quad \text{and} \quad \Delta E_Q = a\Sigma - b\Pi + c,$$

where a, b, and c are constants.

The orders of the bonding parameters of a number of complexes correlate well with the sum of the Hammet $\Sigma_i\sigma_i$ constants of the peripherally attached substituents.

Nevertheless the influence of substituents on the bonding properties of the macrocyclic Pc ring to the central iron atom is not completely understood, and some complexes [e.g., $(C_6H_5S)_4PcFe$] do not fit into this formalism.

A tetra-substituted PcFe having four iron porphyrin units attached has been published [47]. The spectrum (measured at 4.2 K) was evaluated as a sum of two components. The narrow doublet ($\delta = 0.37$ mm/s, $\Delta E_Q = 0.88$ mm/s) corresponds to the iron atoms in the porphyrin moiety, the second doublet ($\delta = 0.44$ mm/s, $\Delta E_Q = 2.57$ mm/s) to those in the phthalocyanine macrocycle.

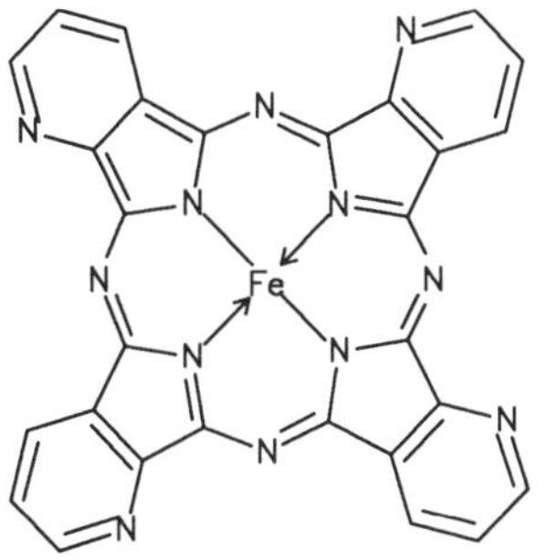

Figure 5 Structure of the T(2,3-Py)PFe.

The hyperfine parameters of the aza-substituted T(2,3-Py)PFe (Fig. 5) are similar to the octacyano- or hexadecachloro-substituted phthalocyaninatoiron macrocycles. This may be interpreted as a result of the electron-withdrawing behavior of the ring-nitrogen atoms.

Table 1 Mössbauer Data of PcFe and Substituted R_xPcFe Macrocycles

Complex	T (K)	δ^a (mm/s)	ΔE_Q (mm/s)	Ref.
PcFe	RT	0.49^b	2.62	[1]
	4	0.49	+2.70	[19]
	77	0.51	+2.69	
	293	0.40	+2.62	
	4.8	0.58	2.70	[21]
	81	0.64	2.64	
	293	0.66	2.64	
	85	0.42	2.61	[22]
	140	0.46	2.63	
	164	0.45	2.61	
	190	0.45	2.59	
	250	0.40	2.60	
	300	0.38	2.57	
	293	0.38	2.58	[35]
	77	0.49	2.46	[32]
	297	0.41	2.62	
	77	0.28	2.73	[32]
	297	0.19	2.74	
	77	0.77^c	2.71	[48]
	195	0.74^c	2.65	
	295	0.68^c	2.67	
	423	0.60^c	2.58	
	4.6	0.81^c	2.61	[49]
	77	0.79^c	2.62	
	295	0.72^c	2.60	
	RT	0.63^c	2.62	[50]
$(CH_3)_8$PcFe	77	0.47	2.61	[35]
	293	0.38	2.55	
	5.1	0.49	2.66	[46]
	79	0.48	2.67	
	297	0.38	2.58	
$(C_2H_5)_4$PcFe	293	0.38	2.57	[51]
$(C_8H_{17})_8$PcFe	83	0.45	2.58	[52]
$[(CH_3)_3C]_4$PcFe	79	0.43	2.42	[46]
$(CH_3O)_8$PcFe	293	0.36	2.50	[35]
	4.2	0.48	2.54	[46]
	79	0.46	2.53	
	297	0.33	2.49	
$(OCH_2O)_4$PcFe	4.6	0.47	2.88	[46]
	79	0.47	2.80	
	292	0.37	2.61	

(continued)

Table 1 *(continued)*

[(CH₃)₃CCH₂O]₄PcFe	4.2	0.28	2.91	[46]
	79	0.28	2.91	
	299	0.20	2.53	
(C₆H₅S)₄PcFe	4.4	0.24	1.38	[46]
	79	0.23	1.37	
	296	0.14	1.36	
(HOOC)₄PcFe	77	0.17	1.24	[44]
		0.56	0.82	
		0.30	2.92	
	195	0.13	1.23	
		0.52	0.83	
		0.25	2.91	
(HOOC)₈PcFe	77	0.24	2.55	[44]
		0.47	0.83	
	195	0.20	2.56	
		0.42	0.82	
	293	0.14	2.50	
		0.35	0.79	
[3,4-(COOH)₂C₆H₃C(O)]₂PcFe	77	0.05	2.01	[44]
		0.48	0.72	
	195	0.02	2.03	
		0.43	0.73	
(C₁₀H₂₁OOC)₄PcFe	4.3	0.29	2.78	[45]
	78.8	0.29	2.71	
	298	0.22	2.61	
(C₁₀H₂₁OOC)₈PcFe	4.2	0.23	2.48	[45]
	79.3	0.20	2.50	
	298	0.17	2.48	
[C(O)N(C₁₀H₂₁)C(O)]₄ PcFe	4.3	0.30	1.96	[45]
	79.3	0.27	1.96	
	298	0.16	1.96	
(CN)₄PcFe	293	0.36	0.67	[53]
		0.14	1.47	
(CN)₈PcFe	4.3	0.27	1.56	[45]
	79.0	0.27	1.56	
	298	0.20	1.56	
Cl₁₆PcFe	4.5	0.32	1.72	[45]
	78.7	0.32	1.65	
	291	0.25	1.69	
(NH₂)₄PcFe	4.5	0.27	2.26	[43]
		0.44	0.78	
	293	0.37	0.71	[54]
(NO₂)₄PcFe	4.3	0.22	1.99	[45]
	78.5	0.22	1.99	
	298	0.15	2.00	

(continued)

Table 1 *(continued)*

	4.5	0.22 0.39	1.99 0.67	[43]
TsPcFe[d]	77	0.46[c] 0.12[c]	1.46 1.31	[55]
	195	0.40[c] 0.07[c]	1.46 1.28	
	298	0.29[c] 0.10[c]	1.47 1.09	
TsPcFe[e]		-0.03[b]	1.54	[56]
TsPcFe[f]	293	0.34	2.42	[57]
T(2,3-Py)PFe[g]	293	0.23	1.63	[58]

[a] Relative to metallic iron, otherwise noted.
[b] Stainless steel reference.
[c] Sodium nitroprusside reference.
[d] Reported for a compound "$Na_3FeSPc \cdot 0.5 \ NaOH \cdot 4 \ H_2O$".
[e] Prepared according to Weber and Busch [59].
[f] Prepared according to A. Hirsch and M. Hanack [57].
[g] A second doublet was assigned to $[T(2,3-Py)PFe]_2O$ ($\delta = 0.34$ mm/s, $\Delta E_Q = 0.83$ mm/s).

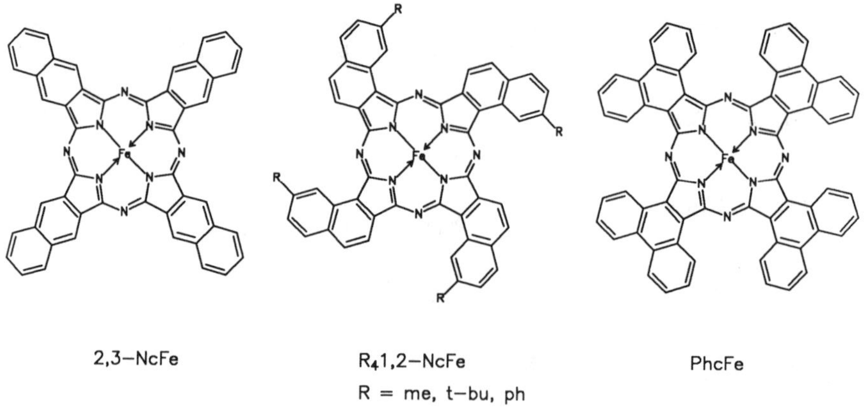

2,3−NcFe   R₄1,2−NcFe   PhcFe
                       R = me, t−bu, ph

Figure 6 Structures of 2,3- and R_4-1,2-NcFe(II) (one isomer, C_{4h} symmetry) and PhcFe(II) macrocycles.

A number of phthalocyaninatoiron complexes fused with one or two additional benzene rings (Fig. 6), optionally substituted, have been prepared and characterized [60−64]. The isomer shift values are nearly the same as those for PcFe(II), and the quadrupole splittings are in the range of $\Delta E_Q = 2.21 - 2.69$ mm/s (Table 2).

Table 2 Mössbauer Data of 2,3- and 1,2-NcFe(II) and PhcFe(II) Complexes

Complex	T (K)	δ (mm/s)[a]	ΔE_Q (mm/s)	Ref.
2,3-NcFe	293	0.36	2.21	[61]
	4.3	0.47	2.36	[46]
	79	0.46	2.31	
	299	0.37	2.28	
1,2-NcFe	293	0.35	2.59	[63]
$(CH_3)_4$-1,2-NcFe	293	0.36	2.61	[62]
$((CH_3)_3C)_4$-1,2-NcFe	293	0.35	2.51	[62]
$(C_6H_5)_4$-1,2-NcFe	293	0.36	2.69	[62]
PhcFe	293	0.36	2.43	[64]

[a] Relative to metallic iron.

ii. Measurements on Hexacoordinated PcFe(II)L$_2$, PcFe(II)LL', and Bridged [PcFe(II)L]$_n$ Complexes

A great variety of hexacoordinated PcFe(II) complexes has been synthesized. The reaction of PcFe(II) with unidentate ligands yields PcFeL$_2$ complexes. With bidentate ligands monomeric complexes PcFeL$_2$ as well as bridged complexes [PcFeL]$_n$ can be synthesized. The bridged complexes [PcFeL]$_n$ were studied because of their special semiconducting, magnetic, and optical properties [3, 4]. We have reported details of the synthesis and general properties of these compounds [4, 5]. In addition monomeric complexes PcFe(II)LL' with two different ligands L and L' have been described [65].

Although only a few x-ray investigations of hexacoordinated complexes have been carried out, the two ligands are generally assumed to occupy perpendicular positions, for example, in the complex PcFe(4-mepy)$_2$ [66].

Most probably the electron configuration [39] of the ground state of PcFe(II) is $(d_{xy})^2 (d_{xz}, d_{yz})^3 (d_{z^2})^1$. The ligand molecules form σ-bonds with the a_{1g} orbitals of the central iron atom. If the ligand also contains suitable π orbitals, π-type bonds with the corresponding e_g orbitals of the iron atom will also be formed. The resulting hexacoordinated complexes PcFeL$_2$ and [PcFeL]$_n$ are generally low-spin iron(II) systems. The changes in the hyperfine parameters caused by the axial ligation were the subject of many investigations.

As a result of the increased s-electron density at the iron nucleus caused by the σ bonding, the isomer shift of the hexacoordinated complexes decreases with respect to PcFe. A change in electron density in the e_g orbitals by the π interaction also influences the isomer shift, though to a smaller extent. With increasing π-acceptor ability of the axial ligands, reduced shielding by the $3d$ electrons gives rise to a decrease in the isomer shift.

Neglecting the lattice contribution, the field gradient can be associated with the occupancy of the p and d orbitals. The valence contribution to the EFG can be expressed in terms of the effective population of the d orbitals [67]:

$$V_{zz}/e = {}^4/_7(1-R) <r^{-3}> \{n(d_{x^2-y^2}) - n(d_{z^2}) + n(d_{xy})$$
$$- {}^1/_2 [n(d_{xz}) + n(d_{yz})]\},$$

where V_{zz} is the principal component of the electric field gradient, R the Sternheimer antishielding factor, $n(d_i)$ the effective population of the various $3d$ orbitals, and $<r^{-3}>$ the expectation value of r^{-3} for the radial part of the $3d$ wave function.

Both an enhanced electron density in those orbitals that contribute negatively to V_{zz} (d_{z^2}, d_{xz}, d_{yz}) and a diminished electron density in the $d_{x^2-y^2}$ orbital will decrease $|\Delta E_Q|$. Therefore $|\Delta E_Q|$ is expected to increase with an increasing π-acceptor strength of the axial ligands.

Assuming a local D_{4h} symmetry around the iron center, the following general and qualitative predictions concerning the influence of the σ-donor and π-acceptor strength of the axial ligands L to the hyperfine parameters were made [68]:

1. δ would be expected to decrease both with increasing σ-donor strength and with an increasing π-acceptor ability of the axial ligands L.
2. ΔE_Q would be expected to decrease with increasing σ-donor strength but to increase with increasing π-acceptor strength of the axial ligands L.

These predictions were made on the basis of a magnetic field study, which proved a positive sign for V_{zz} for the complex PcFe(py)$_2$ [68]. Measurements in an external magnetic field have been carried out for only a few complexes. For the complexes PcFeL$_2$ (L = pip [69], mesNC [35]) and PcFe(CO)MeOH [69], positive values of V$_{zz}$ were reported.

These predictions, however, are only valid if complexes with similar Fe−N (equatorial) and Fe−L (axial) bond lengths are compared, as will become apparent later.

Most of the complexes PcFe(II)LL' and [PcFe(II)L]$_n$ reported in the literature can be assigned to one of two groups having different Mössbauer parameters. In addition to the complexes with L = N-, S- or P-donor ligands the carbonyl complexes PcFe(CO)L and the complexes PcFeL$_2$ and [PcFeL]$_n$ with L = isocyanides may be distinguished. The Mössbauer parameters for the monomeric complexes PcFeL$_2$ with L = N-, S- or P-donor ligands are listed in Table 3. Since complexes with O-donor ligands are very unstable, no Mössbauer data of such species are yet available. Monomeric complexes with bidentate ligands are listed in Table 4.

The isomer shift values of complexes with N- and S-donor ligands are decreased with respect to PcFe. δ varies very little ($\delta = 0.22 - 0.36$ mm/s), indicating that the s-electron density is not strongly dependent on the nature of the bonds to nitrogen or sulfur in the different complexes [70]. The complexes with P-donor ligands show lower isomer shifts ($\delta = 0.16 - 0.13$ mm/s). This may be explained by the enhanced σ-donor strength as well as by the π-acceptor capacity of these ligands [71].

These simple considerations fail to explain the changes in quadrupole splitting. The aliphatic amines as axial ligands have no π-acceptor ability. Tertiary amines have an increased σ-donor strength in comparison to secondary and primary amines. An enlarged population in the d_{z^2} orbital should cause lower values of ΔE_Q. However, measurements show that ΔE_Q of the PcFeL$_2$ complexes increases in the following order (for values see Tables 3 and 4):

NH_3 < n-butylamine, propylamine, ethylendiamine < diethylamine, piperidine < dabco.

An explanation for this behavior seems to be the increasing steric requirement of the alkyl groups attached to nitrogen. Elongated axial Fe$-$N bonds and a poor overlap of the lone pair N orbital with the iron $4s$ and $3d_{z^2}$ orbitals will cause the value of ΔE_Q to increase. A further indication that these considerations are correct is the conspicuously enlarged value of δ for PcFe(dabco)$_2$ and [PcFe(dabco)]$_n$, probably caused by the slight σ-donating effect of dabco to the iron $4s$ orbital (Table 4).

It is remarkable that the monomeric PcFe(dabco)$_2$ and the bridged [PcFe(dabco)]$_n$ complexes show a quadrupole splitting (about 2.9 mm/s) even greater than PcFe (about 2.6 mm/s). As explained in Chapter D, PcFe in its ß modification has to be considered not as purely square planar, since the nitrogen atoms of the neighboring molecules are situated axially above and below the central iron atom at a distance of about 338 pm. The reason for the enlarged ΔE_Q in the dabco complexes must be the very weak interaction of the σ-donating nitrogen atom of the dabco ligand, comparable to the weak interaction of the central iron atom in ß-PcFe with the neighboring macrocycle. The axial bonding may also cause changes in the square-planar bonding of the macrocycle to the central iron atom, as will be discussed in the following. However, due to the lack of crystallographic data for PcFe(dabco)$_2$, it is not possible to explain these effects satisfactorily.

Comparatively high values of ΔE_Q are also observed in the piperidine complexes of substituted phthalocyanines as in $(t$-bu)$_4$PcFe(pip)$_2$, T(2,3-Py)PFe(pip)$_2$, and in the corresponding TPPFe(pip)$_2$ [72a] complex, the crystal structure of which shows comparatively long axial bond distances [72b].

Similar problems arise when discussing the effects of axial ligands that can form π interactions with the iron e_g orbitals. For example, the contrary influences of a stronger σ-donor and π-acceptor ability on the value of ΔE_Q going from PcFe(dabco)$_2$ to PcFe$\{$(et)$_3$P$\}_2$ cannot be easily explained.

Table 3 Mössbauer Parameters of the Complexes PcFeL$_2$ with Unidentate Ligands (L = N-, S-, or P-donor Ligand)

Complex	T (K)	δ^a (mm/s)	ΔE_Q (mm/s)	Ref.
PcFe(NH$_3$)$_2$	293	0.26	1.79	[73]
PcFe(n-C$_3$H$_7$NH$_2$)$_2$		0.50^b	1.97	[74]
PcFe(n-buNH$_2$)$_2$	77	0.34	1.94	[75]
	293	0.25	1.87	[76]
TsPcFe(n-C$_8$H$_{17}$NH$_2$)$_2{}^c$	293	0.24	1.89	[57]
PcFe(HOCH$_2$-CH$_2$-NH$_2$)$_2$·2THF		0.53^b	2.01	[74]
PcFe((et)$_2$NH)$_2$		0.51^b	2.22	[74]
PcFe(pip)$_2$	4.2	0.34	+2.24	[69]
	115	0.36	2.25	
	295	0.28	2.34	

(continued)

Table 3 *(continued)*

	77	0.33	2.21	[75]
	77	0.36	2.19	[70]
PcFe(pip)$_2$·2THF		0.52^b	2.25	[74]
(t-bu)$_4$PcFe(pip)$_2$^{c}	293	0.27	2.14	[57]
T(2,3-Py)PFe(pip)$_2$	293	0.27	2.26	[58]
PcFe(C$_6$H$_5$NH$_2$)$_2$		0.53^b	2.11	[74]
PcFe(C$_6$H$_5$CH$_2$NH$_2$)$_2$·2THF		0.55^b	1.53	[74]
PcFe(py)$_2$	77	0.14^d	1.89	[79]
	295	0.06^d	2.04	
	4.2	0.32	+1.96	[68]
	77	0.62^b	1.89	[48]
	195	0.58^b	2.00	
	295	0.53^b	2.05	
	77	0.33	1.97	[75]
	293	0.26	2.02	
	4.6	0.70^b	1.92	[49]
	77	0.62^b	1.96	
	295	0.63^b	2.07	
		0.52^b	1.99	[74]
	77	0.32	1.94	[70]
	298	0.20	2.06	[77]
PcFe(py)$_2$^{e}	77	0.62^b	1.94	[49]
me$_8$PcFe(py)$_2$	293	0.26	1.97	[35]
(meO)$_8$PcFe(py)$_2$	293	0.25	1.95	[35]
TsPcFe(py)$_2$^{c}	293	0.22	2.00	[57]
PcFe(2-mepy)$_2$	195	0.60^b	1.94	[48]
	295	0.56^b	2.01	
	293	0.25	1.99	[78]
PcFe(3-mepy)$_2$	77	0.61^b	1.81	[48]
	195	0.60^b	1.88	
	295	0.55^b	1.95	
	77	0.32	1.87	[70]
PcFe(4-mepy)$_2$	195	0.58^b	1.89	[48]
	295	0.55^b	1.96	
	77	0.35	1.97	[70]
PcFe(4-t-bu-py)$_2$	293	0.23	1.89	[76]

(continued)

Table 3 *(continued)*

PcFe(3,4-me$_2$py)$_2$	77	0.34	1.98	[70]
PcFe(3,5-me$_2$py)$_2$	77	0.34	1.95	[70]
PcFe(3-HO-py)$_2$	77	0.36	1.90	[70]
PcFe(4-HO-py)$_2$	77	0.27	1.80	[70]
PcFe(3-Cl-py)$_2$	77	0.34	1.91	[70]
PcFe(4-Cl-py)$_2$	77	0.39	2.45	[70]
PcFe(3,5-Cl$_2$-py)$_2$	77	0.36	2.05	[70]
PcFe(py-3-CHO)$_2$	77	0.35	1.97	[70]
PcFe(py-4-CHO)$_2$	77	0.33	1.84	[70]
PcFe(py-3-CN)$_2$	77	0.36	2.17	[70]
PcFe(py-4-CN)$_2$	77	0.34	1.90	[70]
PcFe(im)$_2$	77 195 295	0.63[b] 0.59[b] 0.54[b]	1.71 1.76 1.79	[48]
	77	0.29	1.75	[75]
	77	0.31	1.74	[70]
PcFe(im)$_2$·3THF		0.52[b]	1.73	[74]
PcFe[(et)$_3$P]$_2$	78.6 291	0.25 0.16	1.47 1.54	[71]
PcFe[(n-bu)$_3$P]$_2$	4.3 78.8 291	0.23 0.24 0.15	1.45 1.47 1.57	[71]
PcFe[(etO)$_3$P]$_2$	4.3 78.8 291	0.18 0.17 0.13	0.95 0.99 1.07	[71]
PcFe(THT)$_2$	RT	0.53[b]	2.20	[50]
PcFe(DMSO)$_2$	RT	0.50[b]	2.08	[50]

[a] Relative to metallic iron, otherwise noted.
[b] Sodium nitroprusside reference.
[c] Additional doublet of the uncoordinated macrocycle is present.
[d] No reference given.
[e] In frozen pyridine solution.

The first Mössbauer measurement on PcFe(II)(py)$_2$ with pyridine as the axial ligand was reported in 1966 [79]. Later measurements in the presence of magnetic fields showed that the sign of the electric field gradient at the iron nucleus of the PcFe(II)(py)$_2$ complex is positive [68]. The metal is therefore bonded more strongly to the macrocycle than to the axial ligands [68]. Both the isomer shift and the quadrupole splitting is lowered with respect to PcFe.

The influence of various pyridines substituted with electron-donating or electron-withdrawing groups on the hyperfine parameters of the corresponding complexes was investigated [70]. Although no quantitative interpretation of the data was given, some general trends could be deduced. δ values appear to be correlated with the pK_a and with CT optical absorptions for the complexes with pyridines carrying electron-withdrawing substituents and for those with donor-substituted pyridines. However, no marked correlation between δ and ΔE_Q for these complexes could be observed. Pyridines having substituents in the 2-positions do not form stable complexes with PcFe as a result of steric hindrance [70]. However, the synthesis and the hyperfine parameters of PcFe(2-mepy)$_2$ have been reported [48, 78]. Substituents at the macrocyclic ring have a minor influence on the hyperfine parameters of the corresponding bis-pyridine complexes; however, only a few data of those complexes are available.

The reaction of PcFe with an excess of a bidentate ligand such as pyrazine yields the monomeric complex PcFe(pyz)$_2$. This complex readily splits off pyz in solvents to form a bridged polymer [PcFe(pyz)]$_n$, whereby its one-dimensional structure is responsible for its special electrical and optical properties [3, 4, 5]. The electrical conductivity of the bridged complexes may be enlarged by doping the polymer electrochemically or with an oxidant such as iodine. Due to steric hindrance a number of substituted pyrazines as mepyz, me$_2$pyz, Clpyz, and etpyz form only monomeric complexes. A summary of the data is given in Table 4.

Table 4 Mössbauer Parameters of Monomeric and Polymeric Complexes of PcFe with Bi- and Tridentate Ligands

Complex	T (K)	δ^a (mm/s)	ΔE_Q (mm/s)	Ref.
PcFe(H$_2$N-C$_2$H$_4$-NH$_2$)$_2$·2 C$_2$H$_5$OH	RT	0.26	1.99	[80]
[PcFe(H$_2$N-C$_{12}$H$_{24}$-NH$_2$)]$_n$	293	0.27	1.94	[81]
PcFe(dabco)$_2$	298	0.59^b	2.89	[82]
	293	0.32	2.91	[83]
[PcFe(dabco)]$_n$	298	0.60^b	2.93	[82]
PcFe(pyz)$_2$	298	0.50^b	2.01	[84]
	300	0.24	2.02	[85]
PcFe(2-mepyz)$_2$	298	0.50^b	1.90	[82]
PcFe(2,6-me$_2$pyz)$_2$	298	0.50^b	1.97	[82]
	293	0.24	1.97	[76]
PcFe(Clpyz)$_2$	298	0.51^b	2.15	[82]
PcFe(etpyz)$_2$	298	0.50^b	2.02	[82]
[PcFe(pyz)]$_n$	300	0.24	2.03	[85]
	83	0.31	1.85	
	298	0.18	1.98	[77]

(continued)

Table 4 *(continued)*

$[PcFe(pyz) \cdot 0.3 \ C_6H_5Cl]_n$	298	0.50[b]	2.05	[84]
$[PcFe(pyz) \cdot 0.5 \ C_6H_6]_n$	298	0.50[b]	2.01	[84]
$[PcFe(pyz)I_{0.38}]_n$	298	0.50[b]	2.05	[84]
$[PcFe(pyz)I_{0.77}]_n$	298 4.2	0.50[b] 0.58[b]	2.04 1.90	[84]
$[PcFe(pyz)I_{1.49}]_n$	298	0.49[b]	2.00	[84]
$[PcFe(pyz)I_{2.54}]_n$	298	0.49[b]	1.99	[84]
$[PcFe(pyz)I_{2.76}]_n$	298	0.50[b]	2.03	[84]
$[PcFe(pyz)I_{0.26}]_n$	300	0.25 0.18	2.05 1.13	[85]
$[PcFe(pyz)(BF_4)_{0.5}]_n$	293	0.49[b]	1.98	[86]
$[(t\text{-bu})_4PcFe(pyz)]_n$	293	0.24	1.99	[87]
$[(meO)_8PcFe(pyz)]_n$	293	0.24	1.87	[88]
$[(C_5H_{11}O)_8PcFe(pyz)]_n$	293	0.27	2.26	[89]
$TsPcFe(pyz)_2$ [c]	293	0.22	1.92	[57]
$[TsPcFe(pyz)]_n$	293	0.24	2.04	[57]
$T(2,3\text{-Py})PFe(pyz)_2$ [d]	293	0.23	1.69	[58]
$[T(2,3\text{-Py})PFe(pyz)]_n$ [d]	293	0.23	1.69	[58]
$[PcFe(bpy)]_n$	300	0.23	1.97	[85]
$[PcFe(bpy)I_{1.06}]_n$	298 300	0.24 0.24 0.20	2.00 1.96 1.05	[77] [85]
$[PcFe(bpe)]_n$	298	0.20	2.01	[77]
$[PcFe(bpa)]_n$	298	0.28	2.04	[77]
$[PcFe(tmbpy)]_n$	298	0.28	2.06	[77]
$PcFe(dt)_2$	293	0.28	2.36	[90]
$PcFe(pym)_2$	293	0.26	1.99	[35]
$PcFe(pdz)_2$	293	0.21	1.82	[35]
$PcFe(taz)_2$	293	0.26	1.95	[35]
$PcFe(tz)_2$[c]	293	0.15	1.79	[61]
$[PcFe(tz)]_n$	293 112	0.13 0.19	2.27 2.16	[61]
$[PcFe(bpytz)]_n$	293	0.24	1.90	[83]
$[me_8PcFe(tz)]_n$	293	0.14	2.11	[91]

(continued)

Table 4 *(continued)*

$[(meO_8)PcFe(tz)]_n$ [c]	293	0.15	2.19	[91]
$PcFe(me_2tz)_2 \cdot 0.5\ me_2tz$	77	0.36	2.67	[35]
$2,3\text{-}NcFe(pyz)_2$	293	0.26	1.90	[60a]
$[2,3\text{-}NcFe(tz) \cdot 0.5\ CHCl_3]_n$	293	0.19	1.97	[61]

[a] Relative to metallic iron, otherwise noted.
[b] Sodium nitroprusside reference.
[c] Additional doublet of the uncoordinated macrocycle is present.
[d] Additional doublet of an impurity is present.

The monomeric complexes of pyrazine and the substituted pyrazines, the polymeric $[PcFe(pyz)]_n$, as well as the bridged complexes of bpy, bpa, bpe, and tmbpy, have almost the same hyperfine parameters as the pyridine complex $PcFe(py)_2$. This confirms the similar σ-donating and π-accepting ability of these ligands.

In the case of the bridged complexes these data support the assumed linear chain structure ("shish kebab" structure) of the complexes $[PcFeL]_n$ containing hexacoordinated iron centers.

Investigations on the effect of doping of the bridged complexes lead to different results. The measurements of $[PcFe(pyz)]_n$ complexes doped with various amounts of iodine as well as doped pyrazine bridged complexes synthesized from iodine-doped monomer $PcFe(pyz)_2$ results in hyperfine parameters nearly identical with the values for the undoped polymer. There is no evidence for a second doublet arising from an oxidized Fe(III) species. These results argue that the polymer structure remains intact upon iodination and imply that the oxidation involves orbitals that are not predominantly metal in character [84]. The investigation of electrochemically doped $[PcFe(pyz)(BF_4)_x]_n$ supports these considerations [86].

In contrast to these results the occurrence of a second doublet assigned to Fe(III) was reported for $[PcFe(pyz)I_{0.26}]_n$ and $[PcFe(bpy)I_{1.06}]_n$ [85]. The values of the hyperfine parameters for the oxidized complexes ($\delta = 0.18$ mm/s, $\Delta E_Q = 1.13$ mm/s; $\delta = 0.20$ mm/s; $\Delta E_Q = 1.05$ mm/s) are very similar to those reported for the μ-oxo(2) dimer [36]. Probably the different method in doping leads to the formation of this oxidized complex.

In another report the existence of Fe(III) in the complexes $[PcFe(pyz)]_n$ was assumed [77]. However, the materials formed by doping of bridged complexes with iodine are not very well characterized. From their Mössbauer data the authors claimed that Fe(III) species were present.

Although not too many data for pyrazine complexes $R_xPcFe(pyz)_2$ and $[R_xPcFe(pyz)]_n$ of substituted phthalocyaninatoiron(II) are known, there are some interesting trends that should be mentioned. The monomeric and polymeric complexes of T(2,3-py)PFe have identical hyperfine parameters. While δ has the same value as in the uncoordinated macrocycle (see Table 4), the quadrupole splitting is slightly increased [58]. For this observation the same reason as in the case of the dabco complexes of PcFe may be discussed, but no crystallographic data are available.

Contrary to these complexes the pyrazine complexes of the TsPcFe macrocycle show slightly different hyperfine parameters [57]. Values for the polymeric complexes,

$[(t\text{-bu})_4\text{PcFe(pyz)}]_n$, $[(\text{meO})_8\text{PcFe(pyz)}]_n$, and $[(\text{C}_5\text{H}_{11}\text{O})_8\text{PcFe(pyz)}]_n$ are also reported (Table 4). While the t-bu-substituted complex has nearly the same parameters with respect to PcFe(pyz)$_2$, the alkoxy-substituted complexes are different. While the meO substitution causes a decrease in ΔE_Q, the longer $\text{C}_5\text{H}_{11}\text{O}$ groups cause an increase. Since both substituents should have the same electronic influence, this effect may be due to the steric requirement of the longer $\text{C}_5\text{H}_{11}\text{O}$ groups.

The aromatic six-membered heterocycles pyrimidine and pyridazine as well as the 1,3,6-triazine only form monomeric complexes PcFeL$_2$ [35]. The steric requirement of the large phthalocyaninato macrocycle is responsible for this behavior. In contrast to these ligands, the 1,2,4,5-tetrazine can act both as mono- and bidentate ligand [61].

The complexes PcFe(pym)$_2$ and PcFe(taz)$_2$ have hyperfine parameters very similar to the values of the bis-pyridine complex PcFe(py)$_2$, demonstrating that an additional nitrogen atom in the 3-position to the coordinating N-atom influences the bonding to the central metal only to a small degree. However, the ligand pyridazine with a second nitrogen atom in the 2-position causes a significant decrease in δ and ΔE_Q of the complex PcFe(pdz)$_2$. This effect is even more pronounced in the PcFe(tz)$_2$ complex. The reasons for this behavior are not completely understood. Using the simple rules mentioned, these changes should be the result of a distinct σ-donor ligand, which has only a poor π-acceptor ability. However, variations in the length of the bonds of the central metal to the ligands may also cause dramatic changes in the hyperfine parameters. For example, the complex PcFe(me$_2$tz)$_2 \cdot 0,5$ me$_2$tz has hyperfine parameters very similar to PcFe, indicating weak bonding to that ligand, comparable to the tertiary amines mentioned. An edge-on coordination of the ligands pdz and tz is unlikely, as has been demonstrated by NMR measurements on PcRu(pdz)$_2$ [92] and PcRu(tz)$_2$ [61].

The s-tetrazine bridged complexes exhibit semiconducting properties without additional doping (about $10^{-2} - 10^{-1}$ S/cm) [3]. The corresponding complexes with pyrazine as bridging ligand show this increase in conductivity only after chemical or electrochemical doping. This extraordinary behavior was explained by the electronic structures of these complexes. According to LCAO calculations, the electronic structure of [PcFe(pyz)]$_n$ as well as [PcFe(tz)]$_n$ may be described by a band model. While the highest occupied crystal orbital is built up by the iron d_{xy} orbitals, the conductivity band is determined by the π^* orbitals of the bridging ligand [93]. Further extended Hückel calculations described the band structure of [PcFe(tz)]$_n$ characterized by a small band gap between occupied crystal orbitals with a high contribution of the $d_{x^2-y^2}$ orbitals of the iron and crystal orbitals mainly determined by the bridging tetrazine [94].

The hyperfine parameters of [PcFe(tz)]$_n$ show an interesting peculiarity. While δ is similar to the monomeric PcFe(tz)$_2$, ΔE_Q is clearly increased in [PcFe(tz)]$_n$. ΔE_Q is even increased in comparison to [PcFe(pyz)]$_n$ or PcFe(py)$_2$. The same behavior was observed with the bridged s-tetrazine complexes of (me)$_8$PcFe, (meO)$_8$PcFe, and 2,3-NcFe [61, 91]. According to the calculations mentioned, charge carriers may be produced by thermal activation of the electrons of the highest occupied band. As the contribution of such delocalized electrons to the occupation of metal-centered d orbitals is smaller, the increase in ΔE_Q may be the result of the narrow band gap in the one-dimensional chain structure of the [MacFe(tz)]$_n$ complexes.

As 3,6-bis(4-pyridyl)-1,2,4,5-tetrazine coordinates via the pyridine nitrogen atoms, the hyperfine parameters of the bridged complex [PcFe(bpytz)]$_n$ are similar to PcFe(py)$_2$ and not to [PcFe(tz)]$_n$ [83]. In addition, only a poor electrical conductivity was measured for that complex.

In contrast to the complexes $PcFeL_2$ and $[PcFeL]_n$ having identical axial ligands L, a number of complexes of the type $PcFe(CO)L$ have been synthesized and characterized by Mössbauer spectroscopy [65] (Table 5). These complexes show values of δ depending slightly on the ligand L, but that are significantly lower than the values in the hexacoordinated $PcFeL_2$ complexes. However, the values of ΔE_Q are significantly dependent on the nature of the ligand L. While a value similar to the $PcFeL_2$ complexes is observed for $PcFe(CO)THF$ ($\Delta E_Q =$ 1.82 mm/s), a value of $\Delta E_Q = 1.02$ mm/s for the complex $PcFe(CO)(NH_3)$ demonstrates dramatic changes in the electronic structure of the iron atom. The same trend was observed for the corresponding iron porphyrin complexes [7]. For $PcFe(CO)(MeOH)$ a positive sign of the principal component of the EFG has been measured [69]. Note that for the related porphyrinatoiron(II) complexes $(TPP)Fe(CO)_2$ and $(OMBP)Fe(CO)_2$ V_{zz} values were reported to be positive in the case of $(TPP)Fe(CO)_2$, but negative for $(OMBP)Fe(CO)_2$. This shows the influence of the macrocycle on the sign of the principal EFG component [95]. The correlation of various parameters of the ligands L (pK_a values, proton affinities, and donor numbers) with the ΔE_Q values gave no linear correlation, but there is a general trend to lower quadrupole splittings with increasing σ-donor power of the ligand L. Though the carbonyl stretching frequency ν_{CO} may be indicative for the extent of the metal $\rightarrow$ CO π^* backbonding, there is no clear correlation between ν_{CO} and the ΔE_Q values [65].

Two possible qualitative explanations for the observed decrease of ΔE_Q of the carbonyl complexes were given. An increase in Pc$\rightarrow$Fe π donation in order to compensate the Fe$\rightarrow$CO backbonding and a concurrent decrease in Pc$\rightarrow$Fe σ donation were assumed [65, 69]. Alternatively a trans effect of the ligand L was discussed. Therefore the reduction of the d_{xz} and d_{yz} electron population by the π bonding to CO should be more than compensated for by the strong σ component increasing the d_{z^2} population [50, 65].

The value of the bis-carbonyl complex $PcFe(CO)_2$ is already comparable to the small quadrupole splittings observed for the isocyanide complexes (see Table 6) and the isoelectronic $[PcFe(CN)_2]^{2-}$ complexes. The crystal structure of the $(PNP)_2[PcFe(CN)_2]$ complex shows a significant decrease of the axial Fe$\rightarrow$ligand bond lengths in comparison to the $PcFe(4\text{-mepy})_2$ complex (Table 8). Therefore the enhanced electron density along the z axis explains both the decrease of δ and the decrease of ΔE_Q relative to the complexes with N-donor ligands. As CN$^-$ and CO are isoelectronic ligands, a similar effect may explain the hyperfine parameters found for the $PcFe(CO)_2$ complex.

Contrary to the CO ligand, the cyanide anion can act as a bidentate ligand for the synthesis of bridged PcFe complexes, useful as semiconducting materials. There is evidence that, depending on the starting material, different types of bridged complexes $[(H_3O^+)PcFe(CN)]_n$ can be synthesized. Starting from $K_2[PcFe(CN)_2]$ a polymer having only one Mössbauer doublet was measured, in agreement with a CN$-$Fe$-$CN$-$Fe$-$CN$-$ arrangement, obtained by a reaction via pentacoordinated intermediates. Both δ and ΔE_Q were increased in comparison to the monomeric complex, as a result of the fact that the σ-donor and π-acceptor ability of a bridging cyanide ligand acts on two iron centers. A spectrum consisting of two quadrupole doublets (Table 5) of the same intensity is measured for the reaction product of PcFe with KCN. One doublet has the same hyperfine parameters as the material mentioned, indicating the presence of a CN$-$Fe$-$CN unit, while the second one has a significantly smaller quadrupole splitting. In accordance with IR data, a dimeric bridged structure $[NC-Fe-CN-Fe-CN]^{3-}$ was proposed for that complex [35], and the smaller doublet was assumed to be caused by the NC$-$Fe$-$CN arrangement.

The NO derivatives of PcFe have hyperfine parameters similar to those of the six-coordinate PcFe(L)CO complexes [74].

Table 5 Mössbauer Data for PcFeCO(L) and Related Complexes

Complex	T [K]	δ (mm/s)[a]	ΔE_Q (mm/s)	Ref.
PcFe(CO)(THF)	295	0.36	1.82	[65]
PcFe(CO)(H$_2$O)	295	0.37	1.75	[65]
PcFe(CO)(OPPh$_3$)	295	0.36	1.69	[65]
PcFe(CO)(HMPT)	295	0.36	1.60	[65]
PcFe(CO)(DMSO)	295	0.36	1.56	[65]
PcFe(CO)(DMF)	295	0.35	1.56	[65]
PcFe(CO)(MeOH)	295	0.37	1.56	[65]
PcFe(CO)(THT)	295	0.38	1.55	[65]
PcFe(CO)(DEA)	295	0.35	1.45	[65]
PcFe(CO)(pip)	295	0.37	1.27	[65]
PcFe(CO)(pip)[b]	4.2	0.17[c]	+1.48	[69]
	115	0.16[c]	1.51	
	295	0.11[c]	1.57	
PcFe(CO)(py)	295	0.37	1.19	[65]
PcFe(CO)(C$_3$H$_5$NH$_2$)	295	0.36	1.11	[65]
PcFe(CO)(NH$_3$)	295	0.38	1.02	[65]
PcFe(CO)$_2$	295	0.36	0.82	[65]
K$_2$[PcFe(CN)$_2$]	77	0.19[c]	0.56	[75]
(PNP)$_2$[PcFe(CN)$_2$]	RT	0.12[c]	0.67	[80]
[(H$_3$O$^+$)PcFe(CN)]$_n$	77[d]	0.28[c]	1.47	[35]
	77[e]	0.29[c]	1.48	[35]
		0.26[c]	0.78	
PcFe(NO)		0.43	1.69	[74]
Li[PcFe(NO)]·4 THF		0.40	1.74	[74]

[a] Sodium nitroprusside reference, otherwise noted.
[b] Probably the complex PcFeCO(MeOH) has been isolated.
[c] Relative to metallic iron.
[d] Starting from K$_2$[PcFe(CN)$_2$].
[e] Starting from PcFe, additional doublet of unreacted PcFe is present.

Various mono- or multifunctional aliphatic or aromatic isocyanides (see Fig. 7) have been used in our group as coordinating ligands to synthesize monomeric $PcFe(R-NC)_2$ or bridged $[PcFe(CN-R-NC)]_n$ complexes. The Mössbauer data for those complexes are listed in Table 6. The isomer shifts as well as the quadrupole splittings in the monomeric and bridged complexes are significantly smaller than the values of the complexes with the N-donor ligands (Fig. 8). The small values of δ can be related to the strong σ-donor ability of the isocyanide ligands, leading to an increase of the s-electron density at the iron nucleus, which is still amplified by the synergetic effect of the π backbonding to the isocyanide ligand.

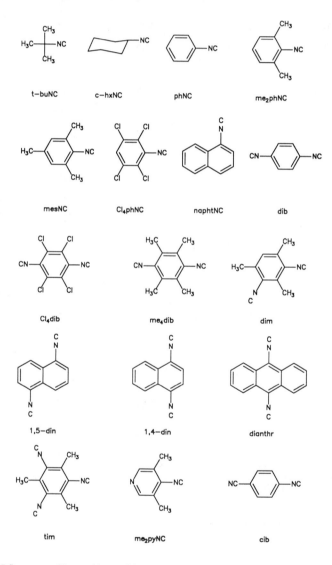

Figure 7 Structures of isocyanides used for the synthesis of $PcFeL_2$ and $[PcFe(L)]_n$ complexes.

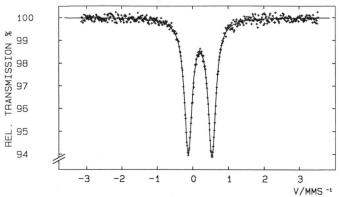

Figure 8 Mössbauer spectrum of PcFe(mesNC)$_2$ (80 K; isomer shift relative to metallic iron).

Table 6 Mössbauer Data of Phthalocyaninatoiron with Isocyanides as Axial Ligands

Complex	T (K)	δ^a (mm/s)	ΔE_Q (mm/s)	Ref.
PcFe(t-buNC)$_2$	293	0.16	0.80	[90]
PcFe(c-hxNC)$_2$	293	0.13	0.69	[90]
PcFe(naphtNC)$_2$	293	0.10	0.65	[96]
PcFe(phNC)$_2$	293	0.11	0.68	[90]
PcFe(me$_2$phNC)$_2$	293	0.12	0.70	[90]
PcFe(Cl$_4$phNC)$_2$	293	0.10	0.68	[90]
PcFe(me$_4$dib)$_2$	293	0.12	0.67	[97]
[PcFe(dib)]$_n$	293	0.12	0.68	[76]
[PcFe(me$_4$dib)]$_n$	293	0.14	0.70	[97]
	80	0.19	0.69	[83]
[PcFe(me$_4$dib)I$_{0.5}$]$_n$	293	0.13	0.69	[90]
[PcFe(me$_4$dib)I$_{1.5}$]$_n$	293	0.10	0.71	[90]
[PcFe(me$_4$dib)I$_{3.0}$]$_n$	293	0.11	0.62	[90]
[PcFe(me$_4$dib)I$_{3.0}$]$_n$	110	0.18	0.57	[90]
[PcFe(Cl$_4$dib)]$_n$	293	0.07	0.75	[97]
PcFe(mesNC)$_2$	293	0.12	0.63	[35]
	80	0.20	0.66	[35]
PcFe(dim)$_2$	293	0.11	0.63	[35]
[PcFe(dim)]$_n$	80	0.20	0.70	[35]
PcFe(tim)$_2$	80	0.19	0.54	[35]

(continued)

Table 6 *(continued)*

[PcFe(tim)$_{0.7}$]$_n$	80	0.19	0.68	[35]
[PcFe(1,5-din)]$_n$	293	0.12	0.64	[35]
[PcFe(1,5-din)I$_{0.95}$]$_n$	293	0.11	0.64	[35]
[PcFe(1,4-din)]$_n$	293	0.11	0.77	[96]
[PcFe(dianthr)]$_n$	293	0.10	0.60	[96]
[me$_8$PcFe(Cl$_4$dib)]$_n$	293	0.07	0.58	[97]
[(et)$_4$PcFe(dib)]$_n$	293	0.11	0.60	[51]
[(et)$_4$PcFe(me$_4$dib)]$_n$	293	0.10	0.64	[51]
(t-bu)$_4$PcFe(t-buNC)$_2$	293	0.13	0.73	[87]
[(t-bu)$_4$PcFe(dib)]$_n$	293	0.09	0.72	[87]
[(t-bu)$_4$PcFe(me$_4$dib)]$_n$	293	0.13	0.72	[87]
[(C$_5$H$_{11}$)$_8$PcFe(dib)]$_n$	293	0.13	0.67	[87]
[(C$_8$H$_{17}$)$_8$PcFe(dib)]$_n$	293	0.10	0.60	[87]
[(2-et-C$_6$H$_{11}$)$_8$PcFe(dib)]$_n$	293	0.11	0.67	[87]
[(C$_5$H$_{11}$O)$_8$PcFe(dib)]$_n$	293	0.12	0.56	[89]
(CN)$_4$PcFe(me$_4$dib)$_2$	293	0.11	0.74	[53b]
[(CN)$_4$PcFe(me$_4$dib)]$_n$	293	0.11	0.72	[53b]
TsPcFe(t-buNC)$_2$	293	0.14	0.82	[57]
TsPcFe(dib)$_2$	293	0.13	0.68	[57]
TsPcFe(me$_4$dib)$_2$	293	0.15	0.91	[57]
[TsPcFe(dib)]$_n$	293	0.15	1.48	[57]
[TsPcFe(me$_4$dib)]$_n$	293	0.14	1.22	[57]
T(2,3-Py)PFe(me$_2$phNC)$_2$	293	0.13	0.69	[58]
[2,3-NcFe(dib)]$_n$	293	0.13	0.58	[60a]

[a] Relative to metallic iron.

While electron-withdrawing substituents at the isocyanide ligand (e.g., in [PcFe(Cl$_4$dib)]$_n$) lead to a further decrease of ΔE_Q in comparison to the unsubstituted complex [PcFe(dib)]$_n$, electron-releasing substituents (e.g., in [PcFe(me$_4$dib)]$_n$) cause an increase. This behavior can be correlated with the IR-stretching frequency of the isocyanide group. The shift of the absorption by changing from free to metal-coordinated ligand is attributed to the σ-donor and π-acceptor abilities of the metal ligand bond. An increase of the π back donation leads to a decrease of the CN-valence frequency since a strong antibonding orbital is available for the π backbonding. In Table 7 data for the complexes $[PcFe(dib)]_n$, $[PcFe(me_4dib)]_n$, and $[PcFe(Cl_4dib)]_n$ are summarized.

Table 7 Isomer Shift Values and IR Valence Frequencies of Isocyanide Complexes $[PcFe(L)]_n$ (L = Cl_4dib, dib, me_4dib)

Ligand L	ν_{CN} free L (cm^{-1})	ν_{CN} coord. L (cm^{-1})	$\Delta\nu_{CN}$ (cm^{-1})	$\Delta\nu_{CN}$ relative to dib (cm^{-1})	δ (mm/s)	δ rel. to $[PcFe(dib)]_n$ (mm/s)
Cl_4dib	2128	2058	-70	-40	0.07	-0.05
dib	2130	2100	-30	0	0.12	0
me_4dib	2113	2092	-21	+9	0.14	+0.02

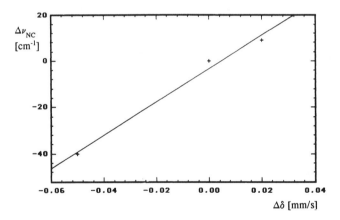

Figure 9 Plot of $\Delta\nu_{NC} = [\nu_{NC}(\text{coord. L}) - \nu_{NC}(\text{free L})] - [\nu_{NC}(\text{coord. dib}) - \nu_{NC}(\text{free dib})]$ versus $\Delta\delta = \delta[(PcFe(L)]_n - \delta[PcFe(dib)]_n$.

A plot of the difference of the change of the stretching frequency versus the difference of the isomer shift indicates a linear correlation (Fig. 9). A similar effect was observed for the complexes $[(t\text{-bu})_4PcFe(L)]_n$ with L = dib and me_4dib, but not for the complexes of TsPcFe. The special properties of the bridged complexes of TsPcFe will be discussed separately.

The δ values of the complexes with the aliphatic isocyanides t-buNC and c-hxNC are slightly increased in comparison to the aromatic isocyanides. As the NC-stretching frequencies in the IR spectrum of complexes with aliphatic isocyanides are shifted to higher wave numbers the π backbonding plays a minor role [90].

Analysis of the ΔE_Q parameters of isocyanide complexes is more complicated and requires a knowledge of the sign of the component of the electric field gradient (EFG) along the z axis as well as information about the corresponding Fe−ligand bond lengths in these complexes. As already mentioned for PcFe, PcFe(py)$_2$, and PcFe(pip)$_2$ [19, 68, 69] positive EFG values have been measured. The Mössbauer spectrum of a microcrystalline sample of PcFe(mesNC)$_2$ recorded at 4.2 K in an external magnetic field of 5 T also demonstrates that in this isocyanide complex the sign of V_{zz} is positive [98].

According to the general rules a decrease of ΔE_Q may be the result of an increased σ-donor power of the axial ligands, while a pronounced π-acceptor behavior, leading to a decrease of the d_{xy} and d_{yz} electron population, should cause an increase of ΔE_Q. As the observed values of the isocyanide complexes are significantly smaller in comparison to the complexes with N-donor ligands, a direct comparison of these groups of ligands becomes very restricted.

The different qualitative explanations for the hyperfine parameters of the PcFe(CO)L and PcFe(PR$_3$)$_2$ complexes may also be applied to the isocyanide complexes. It was assumed that the replacement of axial ligands is accompanied by substantial changes in the bonding between iron and the macrocyclic ligand [71]. Upon replacement of the N-donor ligands by ligands with a better π acceptor ability, there must both be an increase in the π donation from the macrocycle to the iron atom in order to compensate for the iron → ligand backbonding as well as a decrease in the Pc → iron σ donation, which diminishes the electron density in the $d_{x^2-y^2}$ orbital. These phenomena were related to the buffering electronic effect of the conjugated π system of the phthalocyaninato macrocyclic ring [50, 69, 71].

Differences in the axial bond lengths can explain the values of the hyperfine parameters equally well. It was assumed that a shortening of the axial bonds will increase the electron density along the z axis and cause a decrease of ΔE_Q [50]. Structural data now available for 1,2-NcFe(c-hxNC)$_2$ [63] and (t-bu)$_4$-1,2-NcFe(t-buNC)$_2$ [62] confirm these considerations. In contrast to the N-donor complex PcFe(4-mepy)$_2$ [66] in the isocyanide complexes the axial bond lengths are shorter than the equatorial bonds (Table 8). If this applies to all isocyanide complexes, the shortening of the axial Fe−L bonds will increase the s-electron density near the iron d_{z^2}-orbital and result in a decrease of ΔE_Q.

However, the bonding characteristics of the axial ligands have an influence on the equatorial Pc−Fe bonds, which is demonstrated by the small changes in the bond lengths going from PcFe to PcFe(4-mepy)$_2$ and the isocyanide complexes (Table 8).

Table 8 Bond Lengths in the PcFeL$_2$ Skeleton

Complex	Length (pm) Fe−N (eq.)	Length (pm) Fe−L(axial)	Ref.
PcFe	192.7 192.6	−	[99]
PcFe(4-mepy)$_2$	193.7 193.2	204.0	[66]
1,2-NcFe(c-hxNC)$_2$	194.9 193.6	191.1	[63]

(continued)

Table 8 *(continued)*

(*t*-bu)₄-1,2-NcFe(*t*-buNC)₂	192.6 193.3	191.2	[62]
(PNP)₂PcFe(CN)₂	194.2 194.0	196.0	[100]
(PNP)PcFe(CN)₂	193.9 195.7	197.6	[101]
PcFe(DMF)(CO)	193.2 191.2	172.2 Fe−C 207.1 Fe−O	[50]

As demonstrated for the pyrazine bridged polymer, the chemically doped [PcFe(me₄dib)]ₙ and [PcFe(1,5-din)]ₙ retain their polymeric structure, and the oxidation causes only slight changes in the hyperfine parameters of the doped polymers [35, 90].

Substituents at the macrocycle influence the hyperfine parameters, as shown in Table 9 for the bridged complexes [RₓPcFe(L)]ₙ with L = dib and me₄dib. However, no simple correlation of the electron-withdrawing or electron-releasing properties of the substituents with the hyperfine parameters is yet possible. The synthesis of more complexes of this class of bridged complexes will allow an in-depth analysis.

Table 9 Hyperfine Parameters of Substituted Complexes [RₓPcFe(L)]ₙ with L = dib and me₄dib[a]

	PcFe	(et)₄	(*t*-bu)₄	(C₅H₁₁)₈	(C₈H₁₇)₈
dib					
δ (mm/s)	0.12	0.11	0.09	0.13	0.10
ΔE_Q (mm/s)	0.68	0.61	0.72	0.67	0.60
me₄dib					
δ (mm/s)	0.14	0.11	0.13	–	–
ΔE_Q (mm/s)	0.70	0.64	0.72	–	–

	(2-et-C₆H₁₁)₈	(C₅H₁₁O)₈	(CN)₈	TsPcFe	2,3-NcFe
dib					
δ (mm/s)	0.11	0.12	–	0.15	0.14
ΔE_Q (mm/s)	0.67	0.56	–	1.48	0.58
me₄dib					
δ (mm/s)	–	–	0.11	0.14	–
ΔE_Q (mm/s)	–	–	0.72	1.22	–

[a] Measured at 293 K, δ rel. to metallic iron, for ref. see Table 6.

In addition to the electronic properties the steric requirements of the substituents have an influence on the hyperfine parameters; for example, the chain length of an alkyl substituent significantly influences δ and ΔE_Q of the bridged dib complexes [87, 89].

In contrast to the N-donor complexes of TsPcFe, the introduction of sulfonate groups (the complexes contain TsPcFe as trisodium salt [57]) has a conspicuous effect on the hyperfine parameters of the isocyanide complexes. While the complexes with the aliphatic isocyanide t-buNC have very similar hyperfine parameters, the values of δ and ΔE_Q are enlarged for the complex TsPcFe(me$_4$dib)$_2$. Even in comparison to TsPcFe(dib)$_2$ such a change can be detected. Unfortunately no values for PcFe(dib)$_2$ are reported. As mentioned earlier, the influence of the axial ligands must be seen in relation to the equatorial bonding. The low π-acceptor ability of me$_4$dib in combination with the electron-releasing properties of the sulfonate-substituted macrocycle results in hyperfine parameters similar to the complex TsPcFe(t-buNC)$_2$ with the aliphatic isocyanide having inferior π-acceptor strength.

Whereas the isomer shift of the bridged complexes of TsPcFe is not affected very much by the sulfonate groups (see Table 6), the quadrupole splitting ΔE_Q is dramatically enhanced in comparison to the unsubstituted polymers (dib: 0.80 mm/s, me$_4$dib: 0.52 mm/s) and in comparison to the monomeric complexes TsPcFeL$_2$ (L = dib 0.80 mm/s, me$_4$dib 0.31 mm/s). For the PcFe complexes of me$_4$dib, only a slight increase of 0.03 mm/s going from the bisaxially coordinated to the bridged complex was established. Similar behavior (increase of ΔE_Q 0.48 mm/s) was only found for the s-tetrazine coordinated complexes. According to the calculations for [PcFe(tz)]$_n$ [93, 94], the formation of a band structure with a sufficiently small band gap is assumed for these TsPcFe complexes. While in the s-tetrazine complexes the low-lying LUMO of the bridging ligand favors the small band gap, the HOMO of the macrocycle will be the reason in the PcTsFe isocyanide complexes. The determination of the redox potential of TsPcFe complexes [102, 103] demonstrates that as a result of the sulfonate groups the d-orbital energies are higher than those of PcFe.

According to such a band model, the contribution of thermal activated charge carriers to the metal-centered d_{xz} and d_{yz} orbitals is smaller, causing an increased value for ΔE_Q. It is reasonable that an alkyl-substituted ligand has a higher LUMO in comparison to the unsubstituted complex, as has been shown for the dimethyl-substituted s-tetrazine by cyclic voltammetry [104]. As a result of the alkyl substitution, the extent of delocalization in the [TsPcFe(me$_4$dib)]$_n$ is diminished, and a smaller value for ΔE_Q is observed. Though these qualitative considerations are consistent with the experimental data, further work such as molecular-orbital calculations and temperature-dependent Mössbauer measurements on these complexes will be necessary.

As described in previous parts of this chapter, the isocyanide complexes of PcFe have characteristic Mössbauer parameters, clearly distinguishable from the complexes with N bases and from the mixed complexes PcFe(CO)L. Therefore it is very interesting to investigate the coordination chemistry of PcFe with a bidentate ligand having both an isocyanide and a N-base functionality in one molecule. Unfortunately, 4-isocyanopyridine is a very unstable molecule and not suitable for those experiments [105]. However, the methyl-substituted 4-isocyano-3,5-dimethylpyridine is much more stable [35], and it was possible to investigate the coordination behavior of this ambient ligand. The Mössbauer data of the monomeric PcFe(me$_2$pyNC)$_2$ and the bridged [PcFe(me$_2$pyNC)]$_n$ (Table 10) give reliable information about the coordination of the bidentate ligand. The spectrum of the monomer shows only one quadrupole doublet with a small quadrupole splitting, typical for a bis-isocyanide complex. The other possible isomers (see Fig. 10) are not present.

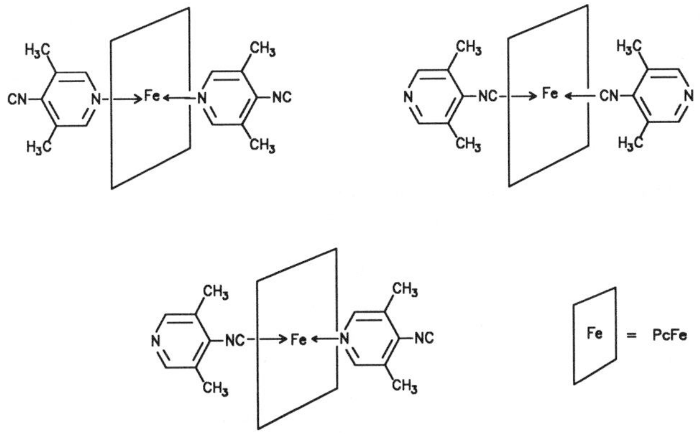

Figure 10 Possible isomers of PcFe(me$_2$pyNC)$_2$.

The bridged polymeric complexes [PcFe(me$_2$pyNC)]$_n$ [35], [(t-bu)$_4$PcFe(me$_2$pyNC)]$_n$ [87], and [TsPcFe(me$_2$pyNC)]$_n$ [57] also show only one quadrupole doublet, indicating that identical iron centers are present. Therefore a regular head-to-tail arrangement of the bridging ligand within the polymer chain as shown in Fig. 11 is plausible. The analysis of ^{1}H-NMR data of the soluble [(t-bu)$_4$PcFe(me$_2$pyNC)]$_n$ provides further evidence for that structure [57]. This regular arrangement supports the polymerization mechanism via reactive pentacoordinated intermediates, which was assumed for the isocyanide complexes [97].

While the isomer shift is nearly the same in the monomeric and polymeric complexes of PcFe with me$_2$pyNC, the quadrupole splitting is enhanced in the polymer. A similar increase was reported for the complex PcFe(CO)py in comparison to PcFe(CO)$_2$ [65]. As ν_{NC} of the bridged complex is diminished, the increase of ΔE_Q may be explained by an increased π backbonding to the ligand.

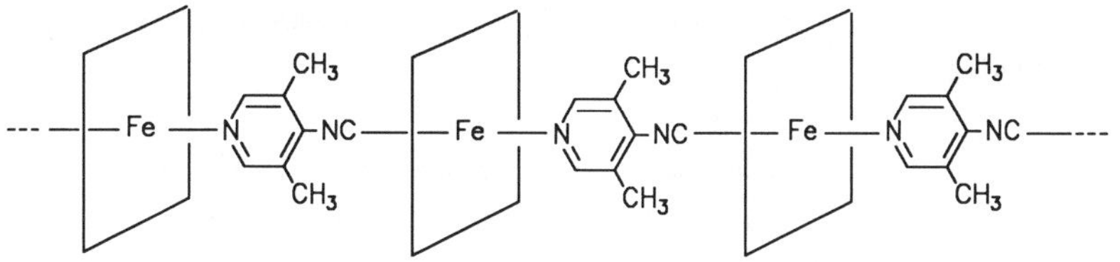

Figure 11 Head-to-tail arrangement of the ambient ligand me$_2$pyNC in the bridged complexes.

A further example for a bidentate ligand is the 4-cyanoisocyanobenzene, which coordinates only via the isocyano group to PcFe [90], having the typical hyperfine parameters of a bis-isocyano complex. Because of this coordination behavior cib does not yield polymeric complexes of PcFe. As the cyano group of cib can coordinate with PcRu, a mixed metal polymer of the type $[PcFe(cib)PcRu(cib)]_n$ is available. As observed for the polymers with me_2pyNC as bidentate ligand, the polymer has the same isomer shift but an enlarged value of quadrupole splitting [35].

Table 10 Mössbauer Data of Complexes with the Ambient Ligands 4-Isocyano-3,5-Dimethylpyridine (me_2pyNC) and 4-Cyanoisocyanobenzene (cib)

Complex	T (K)	δ^a (mm/s)	ΔE_Q (mm/s)	Ref.
$PcFe(me_2pyNC)_2$	293	0.11	0.67	[35]
	88	0.20	0.66	
$[PcFe(me_2pyNC)]_n$	293	0.10	0.91	[35]
	77	0.17	0.87	
$[(t\text{-}bu)_4PcFe(me_2pyNC)]_n$	293	0.12	1.24	[87]
$[TsPcFe(me_2pyNC)]_n$	293	0.13	1.34	[57]
$PcFe(cib)_2$	88	0.20	0.76	[35, 90]
$[PcFe(cib)PcRu(cib)]_n$	293	0.12	1.31	[35]
	88	0.22	1.32	[35]
For comparison:				
$PcFe(me_2phNC)_2$	293	0.12	0.70	[90]
$PcFe(3,5\text{-}me_2py)_2$	77	0.34	1.95	[70]
$PcFe(CO)py$	293	0.11	1.19	[65]

^{a} Relative to metallic iron.

iii. Measurements on Monomeric and Bridged PcFe(III) Complexes and Hexacoordinated Bridged Mixed-Valence Systems

Many penta- or hexacoordinated complexes of PcFe(III) have been described in the literature. Mössbauer spectroscopy was used either to prove the oxidation state and the spin state or to investigate the bonding to axial ligands.

Several chlorophthalocyaninatoiron derivatives and their hyperfine parameters have been described. The reaction of solid PcFe with dry HCl in a nitrogen atmosphere yields an adduct PcFe(II)·4 HCl without oxidation of the metallomacrocycle [106]. Heating of PcFe in nitrobenzene with thionylchloride yields Pc(-I)Fe(+III)Cl$_2$ [107]. Finally, PcFe(III)Cl was prepared by heating of PcFe(II) in concentrated hydrochloric acid under oxidizing conditions [25] or by reacting 1,2-dicyanobenzene with FeCl$_3$ in 1-chloronaphthalene [108].

In addition, several products of the reaction of PcFe with iodine are reported. In these products, for which no further characterization was given, two doublets were found [85]. However, the hyperfine parameters of one doublet are nearly identical with the values of the μ-oxo(1) dimer, which may be formed during the reaction of PcFe with iodine in the presence of oxygen. Other authors reported that the oxidation of PcFe with iodine in 1-chloronaphthalene yields PcFeI and $[PcFeCl]_2I_2$. The structure of $[PcFeCl]_2I_2$ consists of two square-pyramidal PcFeCl units linked by I_2 via the axial chloro ligands. The four halogen atoms form an essentially linear unit. The structure of PcFeI is similar to that of PcNiI [109]. Disordered chains of I_3^- lie in channels between segregated PcFe stacks [110]. The isomer shift values of the halogen complexes PcFeX (X = Cl$^-$, Br$^-$, I$^-$) are identical, and the quadrupole splitting of the bromine and iodine complex is enlarged. The pentacoordinated complexes with X = CF_3COO^-, CCl_3COO^- are in the same range [111]. The Mössbauer data are summarized in Table 11. From magnetic measurements for the PcFeCl complex, an admixed $S = {}^3/_2 / {}^5/_2$ spin state was established, while for the other complexes an $S = {}^3/_2$ spin state with only small admixtures of $S = {}^5/_2$ was deduced [111].

A number of anionic hexacoordinated complexes of the type $B^+[PcFeL_2]^-$ (B = TBA, PNP; L = OH$^-$, OPh$^-$, NCO$^-$, NCS$^-$, and N$_3^-$) show hyperfine parameters (Table 12) that are typical of hexacoordinated low-spin iron(III) in an $S = 1/2$ spin state. For the bis-azide complex $PNP[PcFe(N_3)_2]$ a negative sign of V_{zz} has been published [112]. As in the case of the corresponding complex $(PNP)_2[PcFe(CN)_2]$ (see Table 5), the hyperfine parameters of the $PNP[PcFe(CN)_2]$ complex are conspicuous, and it was mentioned briefly that the applied-field spectra of this complex are different from those of the bis-azide complex. However, the crystal structure demonstrates that the Fe(III) complex has nearly the same bond lengths as the Fe(II) complex [100]. Therefore, the short bonding from Fe → CN may explain the small values for δ and ΔE_Q measured in both bis-cyanide complexes. As observed for the Fe(II) complex, the quadrupole splitting of the polymer $[PcFe(CN)]_n$, containing the cyanide anion in a bridging arrangement, is increased [113].

Table 11 Mössbauer Data of PcFe(III)X Complexes

Complex	T (K)	δ^a (mm/s)	ΔE_Q (mm/s)	Ref.
PcFeCl	77	0.54	2.56	[106]
	77	0.64	2.52	[48]
	195	0.61	2.52	
	295	0.54	2.51	
	298	0.46	2.75	[25]
	111	0.52	2.97	
	RT	0.45	2.71	[110]
	RT	$0.26^{b,c}$	0.31	[114]
	4.8	0.42	2.92	[21]
	81	0.41	2.88	
	222	0.42	2.72	
	4.2	0.28^d	2.94	[111, 115]
	293	0.20^d	2.62	[97]

(continued)

Table 11 *(continued)*

PcFe · 4 HCl	77	0.66	2.33	[106]
	195	0.61	2.31	
	295	0.56	2.33	
PcFeCl$_2$		0.35^b	2.15	[107]
PcFe(p-CH$_3$C$_6$H$_4$SO$_3$)	RT	0.49	3.95	[25]
PcFeBr		0.37	3.16	[74]
	4.2	0.28^d	3.12	[111]
PcFeI$_x$	RT	0.26^d	0.36	[85]
	RT	0.17^d	3.26	
PcFeI	4.2	0.28^d	3.23	[111]
PcFeI$_2$		0.44	3.30	[74]
[PcFeCl]$_2$I$_2$	RT	0.45	2.85	[110]
		0.45	1.02	
PcFe(CF$_3$COO)	4.2	0.28^d	3.08	[111]
PcFe(CCl$_3$COO)	4.2	0.29^d	3.07	[111]
TBA[PcFe(SO$_4$)]^e	4.2	0.55^d	0.82	[111]

^{a} Sodium nitroprusside reference.
^{b} No reference given.
^{c} Probably the spectrum of μ-oxo(1) was measured.
^{d} Relative to metallic iron.
^{e} Additional singlet with δ = 0.61 mm/s is present.

A number of neutral hexacoordinated complexes of the type PcFe(L)CN (L = py, t-bupy, pyz) were synthesized [116−118] beside the anionic and bridged complexes with cyanide. The hyperfine parameters of these complexes are comparable to those of the PcFe(CO)L complexes, demonstrating again the similar coordination behavior of CO and CN⁻ (see Tables 5 and 12). Since the PcFe(CN)L complexes contain a Fe(III) central metal, the diminished occupation of the $3d$ orbitals leads to smaller isomer shift values. The same trend was observed in passing from [PcFe(II)(CN)$_2$]$^{2-}$ to [PcFe(III)(CN)$_2$]⁻.

Table 12 Anionic and Neutral Hexacoordinated PcFe(III) Complexes

Complex	T (K)	δ^a (mm/s)	ΔE_Q (mm/s)	Ref.
PNP[PcFe(OH)$_2$]^b	4.2	0.18	2.22	[111]
PNP[PcFe(OPh)$_2$]^b	4.2	0.22	2.64	[111]
PNP[PcFe(NCO)$_2$]^b	4.2	0.29	2.25	[111]
TBA[PcFe(NCS)$_2$]	4.2	0.26	2.65	[111]

(continued)

Table 12 *(continued)*

PNP[PcFe(N₃)₂]	4.2	0.21	2.49	[111]
	4.2	0.22	-2.47	[112]
PNP[PcFe(CN)₂]	RT	0.03	0.77	[80]
	4.2	0.11	0.99	[111]
PcFe(py)CN	293	0.04	1.57	[116−118]
PcFe(*t*-bupy)CN	293	0.04	1.57	[116−118]
PcFe(pyz)CN	293	0.05	1.60	[116−118]
[PcFe(CN)]ₙ	293	0.09	1.94	[113]

a Relative to metallic iron.
b Additional weak doublet due to μ-oxo complex.

The dimeric complex (py)PcFe(CN)PcFe(NH₃), the trimeric complexes PcFe[PcM(L)CN]₂ (M = Fe, Co, L = py, *t*-bupy), and the polymer {PcFe[PcCo(pyz)CN]}ₙ (see Table 13) are examples of a new class of hexacoordinated bridged mixed-valence systems [116−118]. The central phthalocyanine unit in the trimeric complexes contains iron in the oxidation state +II; the central metal of the terminating phthalocyanine macrocycle is either Fe or Co in the oxidation state +III. With the exception of PcFe[PcFe(*t*-bupy)CN]₂, the hyperfine parameters assigned to Fe(II) are in the typical range observed for low-spin PcFeL₂ complexes (L = N donor; see Table 3), and the iron(III) center in PcFe[PcFe(py)CN]₂ shows almost the same hyperfine parameters as PcFe(py)CN (Table 12), indicating that a charge delocalization between the different metal centers across the bridging cyanide ligand cannot take place in these complexes [116−118]. However, a different behavior was observed for the PcFe[PcFe(*t*-bupy)CN]₂ complex. The isomer shift value assigned to the Fe(II) center of this complex is remarkably small, almost in the range observed for Fe(III) low-spin systems (e.g., [PcFe(CN)]ₙ). A comparatively high increase of the isomer shift (0.17 mm/s) and a decrease of the quadrupole splitting (0.13 mm/s) is observed in the spectrum recorded at 80 K (Fig. 12). These effects make a thermal activation of one electron of the Fe(II) center into a delocalized state at room temperature very likely. The related diminished population of the e_g orbitals of the "Fe(III) site" could readily describe the observed hyperfine parameters. Magnetic measurements of this complex show an unusually high increase of the paramagnetic moment going from 20 to 300 K [76, 117].

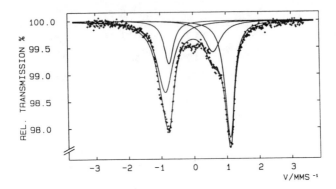

Figure 12 Mössbauer spectrum of PcFe[PcFe(t-bupy)CN]$_2$ [80 K; isomer shift relative to ^{57}Co(Rh) source].

Table 13 Mössbauer Data of Dimeric, Trimeric, and Polymeric Bridged Mixed-Valence Phthalocyaninatometal Complexes

Complex	T (K)		δ^a (mm/s)	ΔE_Q(mm/s)	Ref.
(py)PcFe(CN)PcFe(NH$_3$)	293	Fe(II)	0.26	1.79	[116−118]
		Fe(III)	-0.05	1.47	
PcFe[PcFe(py)CN]$_2$	293	Fe(II)	0.24	2.01	[116−118]
		Fe(III)	0.09	1.61	
PcFe[PcFe(t-bupy)CN]$_2$	293	Fe(II)	0.11	1.98	[116−118]
		Fe(III)	0.03	1.22	
	80	Fe(II)	0.29	1.85	[76]
		Fe(III)	-0.01	1.48	
PcFe[PcCo(py)CN]$_2$	77	Fe(II)	0.33	1.93	[116−118]
PcFe[PcCo(t-bupy)CN]$_2$	293	Fe(II)	0.23	2.01	[116−118]
{PcFe[PcCo(pyz)CN]}$_n$	293	Fe(II)	0.26	2.00	[116−118]

^{a} Relative to metallic iron.

iv. Dimeric μ-Oxo, μ-Nitrido, and μ-Carbido Complexes of PcFe

There has been some uncertainty in the literature about the structure of the μ-oxo dimer of PcFe (see Table 14). The preparation of two different crystalline forms of (μ-oxo)bis(phthalocyaninato)iron complexes was reported [119], depending on the reaction conditions used [described as μ-oxo(1) and μ-oxo(2)]. In fact different species were found in a Mössbauer study, but one doublet was assigned to a sulfonated phthalocyanine [120]. The μ-oxo(1) form was prepared with an alternative synthesis, starting with the PNP[PcFe(OH)$_2$] complex [121]. A second minor component was found having the same Mössbauer parameters as the second product detected in the Mössbauer study mentioned previously [120]. As no sulfuric acid was used in this synthesis, the presence of a sulfonated phthalocyanine can be excluded. The second doublet was assigned to an intermediate hydroxo complex as [PcFe(OH)$_2$]OH [121]. In a careful investigation the existence of the two crystalline forms μ-oxo(1) and μ-oxo(2) was proved [36]. Different samples of μ-oxo(2) were prepared in a number of ways. All samples showed as a main component the so-called μ-oxo(2) dimer and different amounts of minor components. Besides different Mössbauer parameters, the μ-oxo(1) and μ-oxo(2) show different x-ray powder patterns, IR spectra, and magnetic susceptibility data [36]. Both of them are strongly antiferromagnetically coupled Fe(III) high-spin dimeric systems (S = 5/2). While a bent Fe−O−Fe structure is suggested for μ-oxo(1), a linear or quasi linear Fe−O−Fe bond system is deduced from the observed data for μ-oxo(2). In another study similar results were published, naming the two species PcFeoxyg$_1$ and PcFeoxyg$_2$ [122].

The peripherally substituted complexes [(t-bu)$_4$PcFe]$_2$O [76] and [(C$_5$H$_{11}$O)$_8$PcFe]$_2$O [89] show hyperfine parameters comparable to the values obtained for the linear μ-oxo(2). The steric requirements of those bulky substituents at the macrocycle may favor a linear Fe−O−Fe arrangement.

By reaction of [PcFe]$_2$O with nitrogen bases, adducts of the general formula [(L)PcFe]$_2$O (L = 4-mepy, pip, 1-meim, py) are obtained. A six-coordinate structure and a low-spin (S = 1/2) electronic configuration is assigned to Fe(III) in these μ-oxo derivatives. A weak antiferromagnetic coupling between the iron atoms was deduced from magnetic susceptibility data [123]. The sign of V_{zz} was found to be positive for the complex [(4-mepy)PcFe]$_2$O [112]. The values of the isomer shift and the quadrupole splitting of the hexacoordinated complexes are within the ranges $0.17 - 0.20$ and $1.58 - 1.76$ mm/s (at 4.2 K) and are significantly different from those observed for the five-coordinated μ-oxo(1) and μ-oxo(2) complexes but comparable to the six-coordinated PNP[PcFe(OH)$_2$].

Table 14 Mössbauer Data of Dimeric Oxygen Bridged Phthalocyaninatoiron(III) Complexes

Complex	T (K)	δ^a (mm/s)	ΔE_Q (mm/s)	Ref.
[PcFe]$_2$O μ-oxo(1)	110	0.34	0.38	[120]
"sulfonated PcFe"		0.23	1.28	
μ-oxo(1)^b	4.2	0.37	0.44	[111, 121]
	77	0.36	0.44	
	295	0.25	0.42	
μ-oxo(1)	295	0.26	0.40	[36]
μ-oxo(2)^c	4.2	0.26	1.25	
	77	0.25	1.26	
	295	0.18	1.05	
PcFeoxyg$_1$	78	0.34	0.37	[122]
PcFeoxyg$_2$	298	0.17	1.01	

(continued)

Table 14 *(continued)*

$[(t\text{-bu})_4\text{PcFe}]_2\text{O}$	90	0.22	1.38	[76]
	293	0.13	1.31	
$[(\text{C}_5\text{H}_{11}\text{O})_8\text{PcFe}]_2\text{O}$	293	0.18	1.21	[89]
$[(4\text{-mepy})\text{PcFe}]_2\text{O}^c$	4.2	0.20	1.76	[123]
	77	0.19	1.75	
	295	0.12	1.81	
	4.2	0.20	+1.79	[112]
$[(\text{pip})\text{PcFe}]_2\text{O}^c$	4.2	0.19	1.61	[123]
$[(1\text{-meim})\text{PcFe}]_2\text{O}^d$	4.2	0.17	1.58	[123]
$[(\text{py})\text{PcFe}]_2\text{O}^e$	4.2	0.18	1.73	[123]

[a] Relative to metallic iron.
[b] Additional $[\text{PcFe(OH)}_2]\text{OH}$ is present $[\delta = 0.24$ mm/s; $\Delta E_Q = 1.38$ mm/s (4.2 K)].
[c] Additional μ-oxo(1) is present.
[d] Additional PcFe(1-meim)_2 is present.
[e] Additional PcFe(py)_2, μ-oxo(1), and μ-oxo(2) are present.

In addition to these oxygen bridged dimers, a number of μ-nitrido (Table 15) and μ-carbido complexes (Table 16) were prepared and characterized. As only one Mössbauer doublet is observed for the $[\text{PcFe}]_2\text{N}$ complex, equivalent iron sites with an average intermediate oxidation state of 3.5 were assumed [124, 125]. In the oxidized complex $[(\text{PcFe})_2\text{N}](\text{PF}_6)$ the presence of one doublet with a negative isomer shift value corresponds to iron(IV). Similar low values of δ are measured for the nitrogen base adducts of the type $[((\text{L})\text{PcFe})_2\text{N}](\text{PF}_6)$ (L = 4-mepy, py, pip, 1-meim) [124], also indicating iron(IV) centers in these hexacoordinated complexes. Similar hyperfine parameters were reported for the complex $[(\text{PcFe}(\text{CF}_3\text{CO}_2))_2\text{N}](\text{CF}_3\text{CO}_2)$, while for the corresponding bromine complex $[(\text{PcFeBr})_2\text{N}]\text{Br}$ the appearance of a second doublet indicates two slightly different Fe(IV) environments [125].

Table 15 Mössbauer Data of Dimeric μ-Nitrido Complexes of Phthalocyaninatoiron

Complex	T (K)	δ^a (mm/s)	ΔE_Q (mm/s)	Ref.
$[\text{PcFe}]_2\text{N}$	4.2	0.06	1.78	[125]
	77	0.06	1.76	[124]
$[(\text{PcFeBr})_2\text{N}]\text{Br}$	4.2	-0.08	1.74	[125]
		-0.10	2.02	
$[(\text{PcFe}(\text{CF}_3\text{CO}_2))_2\text{N}](\text{CF}_3\text{CO}_2)$	4.2	-0.10	1.82	[125]
$[(\text{PcFe})_2\text{N}]\text{I}_3$	77	-0.11	1.98	[126]
$[((\text{py})\text{PcFe})_2\text{N}]\text{I}_3$	77	-0.11	1.86	[126]
$[(\text{PcFe})_2\text{N}](\text{PF}_6)$	77	-0.10	2.06	[124]
$[((4\text{-mepy})\text{PcFe})_2\text{N}](\text{PF}_6)$	77	-0.10	1.76	[124]

(continued)

Table 15 *(continued)*

[((py)PcFe)$_2$N](PF$_6$)	77	-0.09	1.76	[124]
[((pip)PcFe)$_2$N](PF$_6$)	77	-0.09	1.73	[124]
[((1-meim)PcFe)$_2$N](PF$_6$)	77	-0.09	1.52	[124]

^{a} Relative to metallic iron.

The isomer shift values of the carbido complex [PcFe]$_2$C and the base adducts [(L)PcFe]$_2$C (L = 4-mepy, py, pip, 1-meim) are indicative of the presence of Fe(IV) in these complexes [127]. Very similar hyperfine parameters were reported for a number of carbido complexes containing neutral (py, THF) or anionic axial ligands (F$^-$, OCN$^-$, SCN$^-$) (Table 16) [128]. Even the complexes [(X)PcFe]$_2$C (X = NO$_3^-$, Br$^-$, Cl$^-$) containing the oxidised cation radical Pc(1-) have comparable hyperfine parameters [128]. The exceptional coordination behavior of the CN$^-$ ligand is again demonstrated by the remarkably small quadrupole splitting of the complex (TBA)$_2$[((CN)PcFe)$_2$C]·2.5 CH$_2$Cl$_2$ [128].

Table 16 Mössbauer Data of Dimeric μ-Carbido Complexes of Phthalocyaninatoiron

Complex	T (K)	δ^a (mm/s)	ΔE_Q (mm/s)	Ref.
[PcFe]$_2$C	77	-0.16^b	2.69	[127]
	77	-0.04	2.67	[128]
[(PcFe)$_2$C] · (CH$_3$)$_2$CO	77	-0.10^b	1.46	[127]
[(4-mepy)PcFe]$_2$C	77	0.03^c	1.19	[127]
[(py)PcFe]$_2$C	77	0.01^c	1.16	[127]
[((py)PcFe)$_2$C] · 0.5 py	77	0.01	1.14	[128]
[(pip)PcFe]$_2$C	77	0.01	1.11	[127]
[(1-meim)PcFe]$_2$C	77	0.01	0.94	[127]
[((THF)PcFe)$_2$C] · THF	77	0.00	1.84	[128]
(TBA)$_2$[((F)PcFe)$_2$C] · 5 H$_2$O	77	0.01	1.14	[128]
(TBA)$_2$[((CN)PcFe)$_2$C] · 2.5 CH$_2$Cl$_2$	77	0.01	0.22	[128]
(TBA)$_2$[((OCN)PcFe)$_2$C] · 2 H$_2$O	77	0.03 0.03	0.78 1.17	[128]
(TBA)$_2$[((SCN)PcFe)$_2$C] · 2 H$_2$O	77	0.03	1.18	[128]
[(ClPcFe)$_2$C]Cl$_{1.5}$ · CH$_2$Cl$_2$	77	0.00	1.16	[128]
[(BrPcFe)$_2$C]Br$_4$ · 0.5 EtOH	77	0.00	1.65	[128]
[((NO$_3$)PcFe)$_2$C] · 2 H$_2$O	77	0.01	1.28	[128]

(continued)

Table 16 *(continued)*

a Relative to metallic iron.
b Contains two additional minor doublets.
c Contains one additional minor doublet.

v. Mössbauer Studies on Phthalocyaninatoiron Complexes as Catalysts

The catalytic properties of phthalocyaninatoiron compounds have been studied in a number of investigations. Especially great efforts were made to examine the electrocatalytic properties. Again Mössbauer spectroscopy is very useful for the investigation and characterization of solid materials such as phthalocyaninatoiron supported on carbon.

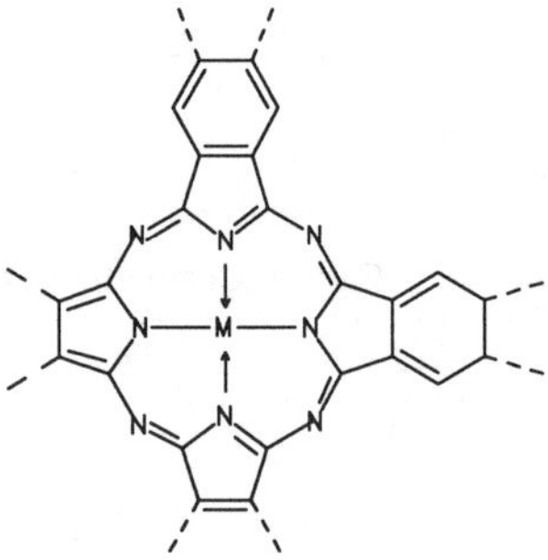

Figure 13 Schematic structure of polyphthalocyanine.

Unfortunately many phthalocyaninato complexes used for catalytic purposes, for example the so-called polyphthalocyanines (Fig. 13), are not very well characterized. Therefore the interpretation of the Mössbauer spectra obtained is difficult and often speculative.

The phthalocyaninatoiron complexes used as catalysts for the oxidation of cumene [129] and of acetaldehyde ethyleneacetal [130] were described by their Mössbauer parameters without further characterization. The Mössbauer data resulting from an investigation of the electrical properties of polyphthalocyaninatoiron are presented without any interpretation [131], and in a further study dealing with this subject, the ratio Fe(II) to Fe(III) is estimated from the intensity data of the Mössbauer spectra [132].

The first study [133] on materials obtained by dissolving PcFe in a coordinating solvent or sulfuric acid, mixing with high surface-area carbon, and precipitating with water, demonstrated the existence of different phthalocyaninatoiron species. The formation of oxidized PcFe was assumed, and was confirmed by oxidation experiments of PcFe with oxygen. It should be noted that commercial-grade PcFe was used, and the doublet assigned to the oxidized PcFe was present in small amounts in the starting material. Using pyridine as a solvent, the presence of the PcFe(py)$_2$ complex was found. In a further investigation [134] using the in situ Mössbauer technique, the species adsorbed on high-surface-area electrodes was examined. In these experiments pyridine was removed by heating after the precipitation of the PcFe(py)$_2$ complex.

By combining electrochemical methods with Mössbauer spectroscopy, it was possible to obtain knowledge about the processes occurring on species of PcFe adsorbed on carbon with regard to the catalytic reduction of oxygen. A series of papers dealt with this topic, using different methods for the preparation of the samples. Besides monomeric PcFe as a starting material [56, 135–137] polyphthalocyaninatoiron was investigated [138–143]. The predominantly very complex spectra were interpreted by assuming an interaction of the surface of the carbon with the iron atoms of the macrocycle. Although the existence of oxidized "PcFe" was mentioned in different articles, neither the structure of this complex nor the structure of the other species was elucidated completely. Not until the structure of the oxygen bridged dimers (PcFe)$_2$O was published [36] could two quadrupole doublets, found in many adsorbed PcFe spectra, be assigned to these complexes [56].

Finally, a Mössbauer study of the systems PcFe/NaY zeolite and PcFe(py)$_2$/NaY zeolite was reported. In order to obtain information about the catalytic mechanism, Mössbauer spectroscopy was used to characterize the state of PcFe inside and outside the zeolite cavities [144].

C. STUDIES ON PHTHALOCYANINATOTIN COMPLEXES

A series of phthalocyaninatotin complexes has been synthesized. The published Mössbauer spectroscopic data for the complexes are listed in Table 17.

Table 17 Mössbauer Data for Phthalocyaninatotin Complexes

Complex	T (K)	δ^a (mm/s)	ΔE_Q (mm/s)	Ref.
PcSn	80.5	+1.41	1.44	[145]
	80	+1.26	1.40	[146]
	82	+1.44	1.43	[147]
	78	+2.80^b	1.43	[148]
Pc$_2$Sn	80	-1.41	0.00	[146]
PcSnF$_2$	80.5	-1.47	0.56	[145]
	80	-1.49	0.70	[146]
PcSnCl$_2$	80.5	-1.17	0.99	[145]
	80	-1.24	0.99	[146]
	78	0.26^b	0.90	[148]
(*t*-bu)$_4$PcSnCl$_2$	78	0.25^b	0.89	[148]

(continued)

Table 17 *(continued)*

PcSnBr$_2$	80.5	-1.19	1.13	[145]
	80	-1.18	1.09	[146]
PcSnI$_2$	80.5	-1.04	1.01	[149]
	80	-1.07	0.99	[146]
	78	0.45[b]	1.21	[148]
PcSn(OH)$_2$	80.5	-1.41	0.47	[145]
	80	-1.43	$-$[c]	[146]
(t-bu)$_4$PcSn(OH)$_2$	78	0.18[b]	0.64	[148]
[(t-bu)$_4$PcSnO]$_n$	78	0.10[b]	0.67	[148]
PcSnFe(CO)$_4$ ^{119}Sn	77	1.32[b]	1.37	[150]
^{57}Fe		-0.10[d]	2.26	

[a] Rel. to PdSn.
[b] Rel. to SnO$_2$.
[c] Splitting not resolved.
[d] Rel. to metallic iron.

The asymmetric environment of the tin atom in PcSn was deduced from the Mössbauer data and confirmed by an x-ray study [147] of the triclinic PcSn(II). The tin atom lies above the plane of the macrocycle, which deviates significantly from planarity. The magnitude of the isomer shift and the lack of quadrupole splitting indicates eight equivalent Sn−N bonds in the square antiprismatic structure of the Pc$_2$Sn(IV) [146].

As a result of the reduced s-electron density at the Sn atom the isomer shifts for the complexes PcSn(IV)X$_2$ (X = F$^-$, Cl$^-$, Br$^-$, I$^-$, and OH$^-$) become more negative with increasing electronegativity of X. There is a linear relation between the measured isomer shifts and the electronegativity of X [146]. The peripheral substitution of the macrocycle by alkyl groups has no appreciable effect on isomer shift and quadrupole splitting [148]. There are no data available for PcSn complexes with electron-withdrawing substituents at the macrocycle.

The temperature dependence of the recoilless fraction f was used to estimate the tightness with which the tin atom is held in the lattice [148]. The polymeric material [(t-bu)$_4$PcSnO]$_n$ [151] shows a significantly different behavior in comparison to the compounds studied, and in particular to its monomeric precursor (t-bu)$_4$PcSn(OH)$_2$. This could be explained by a reduction in vibrational freedom for the central tin atom, resulting from the formation of strong, mostly covalent linkages between PcSn units, forming a co-facially stacked polymer with a PcSn−O backbone [148].

Both ^{119}Sn and ^{57}Fe Mössbauer spectroscopies were used to characterize the oxidation states of the metals in the complex PcSn^{IV}Fe^0(CO)$_4$ [150].

D. STUDIES ON PHTHALOCYANINATOCOBALT COMPLEXES

Whereas the absorption technique uses ^{57}Co in a matrix as a source and the phthalocyaninatoiron complexes as the absorbers, the emission technique uses Pc^{57}Co as source and an iron-containing material as the absorber. As demonstrated in two studies [152, 153] the isomer shift and quadrupole splitting values obtained from the emission spectra of Pc^{57}Co agree reasonably well with those reported for conventional absorption measurements. The electron capture decay of the ^{57}Co to ^{57}Fe therefore causes no disruption of the complex. Emission and absorption spectroscopies were used to get an insight into the nature of the interaction between the central metal atom and the axially located nitrogens of the neighboring molecules in different polymorphs of PcFe and PcCo [33] (Table 18, Fig. 14).

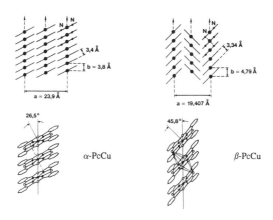

Figure 14 Arrangement of the molecules in the α and ß modification of metal phthalocyanines. [Reproduced, with permission, from D. Wöhrle, G. Meyer, *Kontakte (Darmstadt)*, 3 (1985) 38.]

The larger chemical shift for the ß modification may be explained by a delocalization of π electrons from the neighboring macrocycles onto the $3d_{xz,yz}$ orbitals of the central iron atom. Similar to the coordination of two additional ligands in the complexes PcFeL$_2$, in the ß form the central iron atom in the center of a square of four ligating nitrogens has two additional nitrogen atoms of neighboring molecules in the octahedral positions [154]. However, in the α form the nitrogens of neighboring macrocycles are not in axial positions, and an octahedral arrangement cannot be formed. The larger quadrupole splitting of the ß form may be the result of a not exactly octahedral arrangement of the nitrogen atoms. A slightly asymmetric position may cause a splitting of the $3d_{xy,xz}$ orbitals, which leads to a nonzero value of the asymmetry parameter η.

Table 18 Mössbauer Parameters of the α and ß Modification of Pc^{57}Co and PcFe [33]

	Emission spectra [a] of Pc^{57}Co		Absorption spectra [a] of PcFe	
	δ^b (mm/s)	ΔE_Q (mm/s)	δ^b (mm/s)	ΔE_Q (mm/s)
α form	0.45	2.50	0.46	2.49
	0.45	2.54		
ß form	0.47	2.70	0.48	2.68
	0.47	2.67		

[a] Measured at 80 K.
[b] Relative to metallic iron.

Using the emission spectroscopy technique, ^{57}PcCo, ^{57}PcCo(py)$_2$, and the pentacoordinated ^{57}PcCo(py) complex were investigated [40] (Table 19). Because of the high tendency of PcFe to form the bis-pyridine adduct PcFe(py)$_2$, the corresponding PcFe(py) has never been isolated, but its existence as a reactive intermediate has been discussed [155].

Table 19 Mössbauer Parameters of PcCo, PcCo(py)$_2$, and PcCo(py) Obtained from Emission Spectra [40]

Complex	T (K)	δ^a (mm/s)	ΔE_Q (mm/s)
Pc^{57}Co	77	0.45	2.62
Pc^{57}Co(py)$_2$	77	0.31	2.04
Pc^{57}Co(py)	77	0.46	2.22

[a] Relative to metallic iron.

E. STUDIES ON IODINE CONTAINING PHTHALOCYANINATOMETAL COMPLEXES

A common reagent for oxidizing phthalocyaninatometal complexes is iodine. As a result the conductivity of these partially oxidized phthalocyanines increases considerably. Because ^{129}I Mössbauer spectroscopy is not a standard method, no data for the compounds [MacM(L)I$_x$]$_n$ are so far available. In the complexes PcNiI$_x$ ($x \leq 3$) both ^{129}I Mössbauer measurements and resonance raman spectroscopy indicate that iodine is present as I$_3^-$ [109, 156]. The ^{129}I spectrum is typical of a symmetrical ($D_{\infty h}$) triiodide. There is no evidence of I$^-$ or I$_2$ (see Table 20).

Table 20 ^{129}I Mössbauer Parameters[a]

	δ^b (mm/s)	eQV_{zz} (MHz)^c	Ref.
PcNiI$_{2.16}$	1.34(5)	-1667(10)	[109]
	0.27(5)	-862(10)	
I$^-$	-0.51	0.00	
I$_2$	0.98	1586	

^{a} At 4.2 K.
^{b} Relative to ZnTe.
^{c} Quadrupole coupling constant for ^{129}I.

F. CONCLUSIONS

As has been proved for many other compounds, Mössbauer spectroscopy is a very powerful tool to obtain structural information of various kinds, here mostly for phthalocyaninatoiron systems. From the available Mössbauer parameters (mostly the isomer shift δ and the quadrupole splitting ΔE_Q), sometimes in combination with MO calculations, the electronic structure as well as the spin and oxidation states of the relevant phthalocyaninatometal complexes are ascertained.

In this review, special interest has focused on the coordination behavior of phthalocyaninatoiron and related complexes attached to a whole variety of ligands L. In most of the MacFeL$_2$ monomers and bridged systems [MacFe(L)]$_n$ investigated the structure could not be determined by x-ray methods (many of the compounds could not be obtained as single crystals, and most of the compounds investigated are almost insoluble in organic solvents). Mössbauer spectroscopy is an important method to obtain positive information about the coordination number of the central metal atom, the binding properties between the metal and the ligands, the symmetry of the coordinating ligands, and also, at least in special cases, to obtain some insight into metal$-d$-ligand$-\pi$-orbital interactions.

ABBREVIATIONS

bpa	1,2-bis(4-pyridyl)ethane
bpe	trans-1,2-bis(4-pyridyl)ethene
bpy	4,4'-bipyridine
bpytz	3,6-bis(4-pyridyl)-1,2,4,5-tetrazine
cib	4-cyanoisocyanobenzene
c-hxNC	cyclohexylisocyanide
Cl$_4$phNC	2,3,4,5-tetrachlorophenylisocyanide
dabco	1,4-diazabicyclo[2.2.2]octane
DEA	N,N-diethylformamide
dianthr	9,10-diisocyanoanthracene

(continued)

Abbreviations *(continued)*

dib	1,4-diisocyanobenzene
dim	diisocyanomesitylene
din	diisocyanonaphthalene
DMF	dimethylformamide
DMSO	dimethylsulfoxide
dt	1,4-dithiane
EFG	electric field gradient
H_{int}	internal magnetic field
HMPT	hexamethylphosphorous triamide
I	nuclear spin quantum number
im	imidazole
L, L'	ligands
M	metal
m_I	magnetic nuclear spin quantum number
Mac	macrocycle
MeOH	methanol
meim	methylimidazole
me$_2$phNC	2,6-dimethylphenylisocyanide
me$_2$pyNC	4-isocyano-3,5-dimethylpyridine
mesNC	isocyanomesitylene
naphtNC	1-isocyanonaphthalene
Nc	naphthalocyaninato
OMBP	octamethyltetrabenzoporphyrinato
Pc	phthalocyaninato
pdz	pyridazine
Phc	phenanthrenocyaninato
phNC	phenylisocyanide
pip	piperidine
PNP	bis(triphenylphosphine)nitrogen(1+)
py	pyridine
pym	pyrimidine
pyz	pyrazine
Q	quadrupole moment
R$_x$Pc	peripherally substituted phthalocyanine
RT	room temperature
S	Siemens
taz	1,3,5-triazine
TBA	tetra-*n*-butylammonium
t-bu	tertiary butyl
t-buNC	tert. butylisocyanide
THF	tetrahyrofuran

(continued)

Abbreviations *(continued)*

THT	tetrahydrothiophene
tim	triisocyanomesitylene
tmbpy	4,4'-trimethylenebipyridine
T(2,3-Py)P	tetra(2,3-pyrido)porphyrazinato
TPP	tetraphenylporphyrinato
TsPc	tetrasulfophthalocyaninato
tz	1,2,4,5-tetrazine
V_{ij}	component of the electric field gradient
η	asymmetry parameter
ΔE_Q	quadrupole splitting
δ	isomer shift
ν	stretching frequency
Π	bonding parameter
Σ	bonding parameter

The isomer shifts referred to an iron standard can be converted to a sodium nitroprusside scale by adding +0.257.

The isomer shifts referred to a PdSn standard can be converted to a SnO_2 scale by adding +1.49 [157].

ACKNOWLEDGMENTS

We want to thank Dr. habil. H. Spiering, Johannes Gutenberg Universität Mainz, who put the MOSFUN fitting program at our disposal and gave us much valuable advice. The authors also thank Dr. B. Haas, Dr. W. Hiller and Dr. M. Spohn, Universität Tübingen, for their help in operating the Mössbauer equipment and the computer facilities.

We are grateful to Dr. H. Schultz for helpful discussions and revising this article. We also thank E. Schmid for the help in preparing the manuscript.

REFERENCES

1. L. M. Epstein, *J. Chem. Phys.*, 36 (1962) 2731.

2. M. Hanack, in *Elektrisch leitende Kunststoffe*, H. J. Mair and S. Roth, Eds., 2. Aufl., Hanser, München, 1989.

3. H. Schultz, H. Lehmann, M. Rein, and M. Hanack, in *Structure and Bonding 74*, J. W. Buchler, Ed., Springer Verlag, Berlin, 1990, p.41.

4.   M. Hanack, A. Datz, R. Fay, K. Fischer, U. Keppeler, J. Koch, J. Metz, M. Mezger,
      O. Schneider, and H.-J. Schulze, in *Handbook of Conducting Polymers*, T. A. Skotheim, Ed.,
      Marcel Dekker, New York, 1986, p. 133.

5.   M. Hanack, S. Deger, and A. Lange, *Coord. Chem. Rev.*, 83 (1988) 115.

6.   P. J. Toscano and T. J. Marks, *J. Am. Chem. Soc.*, 108 (1986) 437.

7.   J. R. Sams and T. B. Tsin, in *The Porphyrins*, D. Dolphin, Ed., Academic Press, New York,
      1978.

8.   P. G. Debrunner, in *Physical Bioinorganic Chemistry Series,* Vol. 4, H. B. Gray and
      A. B. P. Lever, Eds., VCH, Weinheim, 1989.

9.   G. J. Long, Ed., *Mössbauer Spectroscopy Applied to Inorganic Chemistry*, Vol. 1, Plenum
      Press, New York, 1984.

10.   U. Gonser, Ed., *Topics Curr. Chem.*, 5 (1975).

11.   A. Vértes, L . Korecz, and K. Burger, *Mössbauer Spectroscopy*, Elsevier, Amsterdam, 1979.

12.   R. H. Herber, Ed., *Chemical Mössbauer Spectroscopy*, Plenum Press, New York, 1984.

13.   N. N. Greenwood and T. C. Gibb, *Mössbauer Spectroscopy*, Chapman and Hall, London, 1971.

14.   D. P. E. Dickson and F. J. Berry, Eds., *Mössbauer Spectroscopy*, Cambridge University Press,
      Cambridge, 1986.

15.   T. E. Cranshaw, B. W. Dale, G. O. Longworth, and C. E. Johnson, *Mössbauer Spectroscopy
      and its Applications*, Cambridge University Press, Cambridge, 1985.

16.   F. J. Berry, in *Physical Methods of Chemistry*, Vol. 4, B. W. Rossiter and J. F. Hamilton,
      Eds., Wiley, New York, 1990, p. 273.

17.   A. R. Champion and H. G. Drickamer, *Proc. Natl. Acad. Sci. U. S. A.*, 58 (1967) 876.

18.   D. C. Grenoble and H. G. Drickamer, *J. Chem. Phys.*, 55 (1971) 1624.

19.   B. W. Dale, R. J. P. Williams, P. R. Edwards, and C. E. Johnson, *J. Chem. Phys.*, 49 (1968)
      3445.

20.   B. W. Dale, *Mol. Phys.*, 28 (1974) 503.

21.   I. Dézsi, A. Balázs, B. Molnár, V. D. Gorobchenko, and I. I. Lukashevich, *J. Inorg. Nucl.
      Chem.*, 31 (1969) 1661.

22.   A. Labarta, E. Molins, X. Vinas, J. Tejada, A. Caubet, and S. Alvarez, *J. Chem. Phys.*, 80
      (1984) 444.

23.   R. A. Stukan, I. S. Kirin, V. Ya. Mishin, and A. B. Kolyadin, *Zh. Neorg. Khim.*, 17 (1972)
      1923.

24.   J. Blomquist and L. C. Moberg, *Phys. Scr.*, 9 (1974) 350.

25.   R. Taube, H. Drevs, E. Fluck, P. Kuhn, and K. F. Brauch, *Z. Anorg. Allg. Chem.*, 364 (1969)
      297.

26.   E. Fluck and R. Taube, *Develop. Appl. Spectrosc.*, 8 (1970) 244.

27.   E. Fluck, Mössbauer and ESCA studies of phthalocyanines, in *Katalyse an Phthalocyaninen*,
      H. Kropf and F. Steinbach, Eds., Thieme, Stuttgart, 1973.

28.   B. Ts. Dudreva and R. K. Pirinchieva, *Bulg. J. Phys.*, 2 (1975) 126.

29.   B. Dudreva, R. Pirintchieva, N. Ivantchev, and S. Grande, *Bulg. J. Phys.*, 3 (1976) 286.

30.   G. Kiss and J. Cirák, M. Hronec, *Proceedings of the International Conference of Mössbauer
      Spectroscopy*, Vol. 1, Bucarest, 1977, p. 273.

31.   G. Kiss, *Petrochemia*, 18 (1978) 55.

32.   M. Hronec and J. Sitek, *React. Kinet. Catal. Lett.*, 21 (1982) 351.

33.   T. S. Srivastava and J. L. Przybylinski, A. Nath, *Inorg. Chem.*, 13 (1974) 1562.

34.   See also experimental section of [110].

35.   A. Lange, Ph.D. Thesis, Universität Tübingen, 1988.

36.   C. Ercolani, M. Gardini, K. S. Murray, G. Pennesi, and G. Rossi, *Inorg. Chem.*, 25 (1986) 3972.

37.   O. Schneider, Ph.D. Thesis, Universität Tübingen, 1983.

38.   E. G. Meloni, L. R. Ocone, and B. P. Block, *Inorg. Chem.*, 6 (1967) 424.

39.   P. Coppens and L. Li, *J. Chem. Phys.*, 81 (1984) 1983.

40.   V. Valenti, P. Fantucci, F. Cariati, G. Micera, M. Petrera, and N. Burriesci, *Inorg. Chim. Acta*, 148 (1988) 191.

41.   R. Behnisch, Ph.D. Thesis, Universität Tübingen, 1989.

42.   J. Simon and P. Bassoul, in *Phthalocyanines*, Volume II, A.B.P. Lever and C.C. Leznoff, Eds., VCH Publishers, Inc., New York, 1992.

43.   N. Kobayashi, H. Konami, T. Ohya, M. Sato, and H. Shirai, *Makromol. Chem., Rapid Commun.*, 10 (1989) 1.

44.   N. Kobayashi, H. Shirai, and N. Hojo, *J. Chem. Soc. Dalton Trans.*, (1984) 2107.

45.   T. Ohya, N. Kobayshi, and M. Sato, *Inorg. Chem.*, 26 (1987) 2506.

46.   T. Ohya, J. Takeda, N. Kobayashi, and M. Sato, *Inorg. Chem.*, 29 (1990) 3734.

47.   N. Kobayashi, Y. Nishiyama, T. Ohya, and M. Sato, *J. Chem. Soc., Chem. Commun.*, (1987) 390.

48.   A. Hudson and H. J. Whitfield, *Inorg. Chem.*, 6 (1967) 1120.

49.   T. H. Moss and A. B. Robinson, *Inorg. Chem.*, 7 (1968) 1692.

50.   F. Calderazzo, G. Pampaloni, D. Vitali, G. Pelizzi, I. Collamati, S. Frediani, and A. M. Serra, *J. Organomet. Chem.*, 191 (1980) 217.

51.   M. Hanack, A. Beck, and K. M. Mangold, *Chem. Ber.*, 124 (1991) 2315.

52.   H. Lehmann, Ph.D. Thesis, Universität Tübingen, 1990.

53.   (a) M. Hanack and R. Grosshans, *Chem. Ber.,* 122 (1989) 1665.
      (b) R. Grosshans, Ph.D. Thesis, Universität Tübingen, 1990.

54.   A. Beck, Ph.D. Thesis, Universität Tübingen, 1990.

55.   J. G. Fanning, G. B. Park, C. G. James, and W. R. Heatley, Jr., *J. Inorg. Nucl. Chem.*, 42 (1980) 343.

56.   A. A. Tanaka, C. Fierro, D. Scherson, and E. B. Yeager, *J. Phys. Chem.*, 91 (1987) 3799.

57.   A. Hirsch and M. Hanack, *Chem. Ber.*, 124 (1991) 833.

58.   R. Thies, Ph.D. Thesis, Universität Tübingen, 1990.

59.   J. H. Weber and D. H. Busch, *Inorg. Chem.*, 4 (1965) 469.

60.   (a) S. Deger, Ph.D. Thesis, Universität Tübingen, 1986.
      (b) M. Hanack and S. Deger, *Isr. J. Chem.*, 27 (1986) 347.

61.   U. Keppeler, S. Deger, A. Lange, and M. Hanack, *Angew. Chem.*, 99 (1987) 349; *Int. Ed. Engl.*, 26 (1987) 344.

62.   M. Hanack, G. Renz, J. Strähle, and S. Schmid, *J. Org. Chem.*, 56 (1991) 3501.

63.   M. Hanack, G. Renz, J. Strähle, and S. Schmid, *Chem. Ber.*, 121 (1988) 1479.

64.   M. Hanack and G. Renz, *Chem. Ber.*, 123 (1990) 1105.

65.   F. Calderazzo, S. Frediani, B. R. James, G. Pampaloni, K. J. Reimer, J. R. Sams, A. M. Serra, and D. Vitali, *Inorg. Chem.*, 21 (1982) 2302.

66.   F. Cariati, F. Morazzoni, and M. Zocchi, *J. Chem. Soc., Dalton Trans.* (1978) 1018.

67.   G. M. Bancroft, M. J. Mays, and B. E. Prater, *J. Chem. Soc. A* (1970) 956.

68.   B. W. Dale, R. J. P. Williams, P. R. Edwards, and C. E. Johnson, *Trans. Faraday Soc.* 64 (1968) 3011.

69.   B. R. James, J. R. Sams, T. B. Tsin, and K. J. Reimer, *J. Chem. Soc., Chem. Comm.* (1978) 746.

70.   G. V. Quédraogo, C. More, Y. Richard, and D. Benlian, *Inorg. Chem.*, 20 (1981) 4387.

71.   T. Ohya, H. Morohoshi, and M. Sato, *Inorg. Chem.*, 23 (1984) 1303.

72.   (a) J. P. Collman, J. L. Hoard, N. Kim, G. Lang, and C. A. Reed, *J. Am. Chem. Soc.*, 97 (1975) 2676.

       (b) L. J. Radonovich, A. Bloom, and J. L. Hoard, *J. Am. Chem. Soc.*, 94 (1972) 2073.

73.   M. Hanack and A. Hirsch, *Synth. Met.*, 29 (1989) F9.

74.   R. Taube, *Pure Appl. Chem.*, 38 (1974) 427.

75.   B. W. Dale, R. J. P. Williams, P. R. Edwards, and C. E. Johnson, *Trans. Faraday Soc.*, 64 (1968) 620.

76.   A. Hirsch, Ph.D. Thesis, Universität Tübingen, 1990.

77.   H.-H. Wei and H.-L. Shyu, *Polyhedron*, 4 (1985) 979.

78.   M. Hanack and P. Vermehren, *Chem. Ber.*, 124 (1991) 1733.

79.   A. Hudson and H. J. Whitfield, *Chem. Commun.* (1966), 606.

80.   W. Kalz, Ph.D. Thesis, Universität Kiel, 1984.

81.   A. Hirsch and B. K. Mandal, unpublished results.

82.   N. K. Jaggi, L. H. Schwartz, and O. Schneider, unpublished results.

83.   A. Lange, unpublished measurement.

84.   B. N. Diel, T. Inabe, N. K. Jaggi, J. W. Lyding, O. Schneider, M. Hanack, C. R. Kannewurf, T. J. Marks, and L. H. Schwartz, *J. Am. Chem. Soc.*, 106 (1984) 3207.

85.   E. Molins, A. Labarta, J. Tejada, A. Caubet, and S. Alvarez, *Transition Met. Chem.*, 8 (1983) 377.

86.   M. Hanack and A. Leverenz, *Synth. Met.*, 22 (1987) 9.

87.   M. Hanack, A. Hirsch, and H. Lehmann, *Angew. Chem.*, 102 (1990) 1499.

88.   O. Schneider and A. Hirsch, unpublished results.

89.   M. Hanack, A. Gül, A. Hirsch, B. K. Mandal, L. R. Subramanian, and E. Witke, *Mol. Cryst. Liq. Cryst.*, 187 (1990) 365.

90.   U. Keppeler, Ph.D. Thesis, Universität Tübingen, 1985.

91.   M. Hanack, A. Lange, and R. Großhans, *Synth. Met.*, 45 (1991) 59.

92.   W. Kobel and M. Hanack, *Inorg. Chem.*, 25 (1986) 103.

93.   E. Canadell and S. Alvarez, *Inorg. Chem.*, 23 (1984) 573.

94.   W. Koch, Ph.D. Thesis, Universität Tübingen, 1986.

95.   K. J. Reimer and C. A. Sibley, *J. Am. Chem. Soc.*, 105 (1983) 5147.

96.   M. Hanack and H. Ryu, *Synth. Met.*, 46 (1992) 113.

97.   U. Keppeler and M. Hanack, *Chem. Ber.*, 119 (1986) 3363.

98.   See [35]; we are indebted to P. Dickmann, Johannes Gutenberg Universität Mainz, for the measurement.

99.   J. F. Kirner, W. Dow, and W. R. Scheidt, *Inorg. Chem.*, 15 (1976) 1685.

100.   H. Küppers, H.-H. Eulert, K.-F. Hesse, W. Kalz, and H. Homborg, *Z. Naturforsch.*, 41b (1986) 44.

101.   H. Küppers, W. Kalz, and H. Homborg, *Acta Crystallogr.*, C 41 (1985) 1420.

102. A. B. P. Lever, S. Licoccia, K. Magnell, P. C. Minor, and B. S. Ramaswan, *Adv. Chem. Ser.*, 201 (1982) 237.

103. A. B. P. Lever, M. R. Hampstesd, G. C. Leznoff, W. Liu, M. Melnik, W. A. Nevin, and P. Seymour, *Pure Appl. Chem.*, 58 (1986) 1467.

104. See [35]; we are indebted to A. Leverenz, Universität Tübingen, for the measurement.

105. J. Koch, Ph.D. Thesis, Universität Tübingen, 1984.

106. L. L. Dickens and J. C. Fanning, *Inorg. Nucl. Chem. Lett.*, 12 (1976) 1.

107. J. F. Myers, G. W. R. Canham, and A. B. P. Lever, *Inorg. Chem.*, 14 (1975) 461.

108. W. Kalz and H. Homborg, *Z. Naturforsch.*, 38b (1983) 470.

109. C. J. Schramm, R. P. Scaringe, D. R. Stojakovic, B. M. Hoffmann, J. A. Ibers, and T. J. Marks, *J. Am. Chem. Soc.*, 102 (1980) 6702.

110. S. M. Palmer, J. L. Stanton, N. K. Jaggi, B. M. Hoffman, J. A. Ibers, and L. H. Schwartz, *Inorg. Chem.*, 24 (1985) 2040.

111. B . J. Kennedy, K. S. Murray, P. R. Zwack, H. Homborg, and W. Kalz, *Inorg. Chem.*, 25 (1986) 2539.

112. E. N. Bakshi and K. S. Murray, *Hyperfine Interact.*, 40 (1988) 283.

113. C. Hedtmann-Rein, Ph.D. Thesis, Universität Tübingen, 1986.

114. H. S. Nalwa and P. Vasudevan, *J. Mater. Sci. Lett.*, 4 (1985) 943.

115. B. J. Kennedy, G. Brain, and K. S. Murray, *Inorg. Chim. Acta*, 81 (1984) L29.

116. A. Hirsch and M. Hanack, in *Opportunities in Electronics, Optoelectronics and Molecular Electronics*, NATO ASI Series E, Vol. 182, J. L. Bredas and R. R. Chance, Eds., Kluwer Academic Publishers, Dordrecht, 1990, p. 163.

117. M. Hanack, A. Hirsch, A. Lange, M. Rein, G. Renz, and P. Vermehren, *J. Mater. Res.*, 6 (1991) 385.

118. M. Hanack, A. Hirsch, A. Lange, M. Rein, G. Renz, and P. Vermehren, *Mat. Res. Soc. Symp. Proc.*, 173 (1990) 189.

119. C. Ercolani, M. Gardini, F. Monacelli, G. Pennesi, and G. Rossi, *Inorg. Chem.*, 22 (1983) 2584.

120. C. S. Frampton and J. Silver, *Inorg. Chim. Acta*, 96 (1985) 187.

121. B. J. Kennedy, K. S. Murray, P. R. Zwack, H. Homborg, and W. Kalz, *Inorg. Chem.*, 24 (1985) 3302.

122. I. Collamati, *Inorg. Chim. Acta*, 124 (1986) 61.

123. C. Ercolani, M. Gardini, K. S. Murray, G. Pennesi, G. Rossi, and P. R. Zwack, *Inorg. Chem.*, 26 (1987) 3539.

124. C. Ercolani, M. Gardini, G. Pennesi, G. Rossi, and U. Russo, *Inorg. Chem.*, 27 (1988) 422.

125. B. J. Kennedy, K. S. Murray, H. Homborg, and W. Kalz, *Inorg. Chim. Acta*, 134 (1987) 19.

126. E. N. Bakshi, C. D. Delfs, K. S. Murray, and H. Homborg, unpublished data, cited in [128].

127. C. Ercolani, M. Gardini, V. L. Goedken, G. Pennesi, G. Rossi, U. Russo, and P. Zanonato, *Inorg. Chem.*, 28 (1989) 3097.

128. E. N. Bakshi, C. D. Delfs, K. S. Murray, B. Peters, and H. Homborg, *Inorg. Chem.*, 27 (1988) 4318.

129. M. Hronec, G. Kiss, and J. Sitek, *J. Chem. Soc., Faraday Trans. I*, 79 (1983) 1091.

130. D. J. Baker, D. R. Boston, and J. C. Bailar, Jr., *Inorg. Nucl. Chem.*, 35 (1973) 153.

131. H. S. Nalwa, *Appl. Organomet. Chem.*, 2 (1988) 257.

132. J. Bijwe, P. S. Pandey, P. Vasudewan, and A. Tripathi, *Eur. Polym. J.*, 23 (1987) 167.

133.   C. A. Melendres, *J. Phys. Chem.*, 84 (1980) 1936.
134.   D. Scherson, S. B. Yao, E. B. Yeager, J. Eldridge, M. E. Kordesch, and R. W. Hoffman, *Appl. Surf. Sci.*, 10 (1982) 325.
135.   A. J. Appleby, J. Fleisch, and M. Savy, *J. Catal.*, 44 (1976) 281.
136.   D. A. Scherson, S. B. Yao, E. B. Yeager, J. Eldridge, M. E. Kordesch, and R. W. Hoffman, *J. Phys. Chem.*, 87 (1983) 932.
137.   D. A. Scherson, C. A. Fierro, D. Tryk, S. L. Gupta, E. B. Yeager, J. Eldridge, and R. W. Hoffman, *J. Electroanal.Chem.*, 184 (1985) 419.
138.   R. Larssson, J. Mrha, and J. Blomquist, *Acta Chem. Scand.*, 26 (1972) 3386.
139.   J. Blomquist, L. C. Moberg, L. Y. Johansson, and R. Larsson, *Inorg. Chim. Acta*, 53 (1981) L39.
140.   J. Blomquist, L. C. Moberg, L. Y. Johansson, and R. Larsson, *J. Inorg. Nucl. Chem.*, 43 (1981) 2287.
141.   J. Blomquist, U. Helgeson, L. C. Moberg, L. Y. Johansson, and R. Larsson, *Electrochim. Acta*, 27 (1982) 1445.
142.   J. Blomquist, U. Helgeson, L. C. Moberg, L. Y. Johansson, and R. Larsson, *Electrochim. Acta*, 27 (1982) 1453.
143.   J. Blomquist, U. Helgeson, L. C. Moberg, L. Y. Johansson, R. Larsson, B. Yom Tov, and L. Jönsson, *Spectros. Lett.*, 18 (1985) 575.
144.   M. Tanaka, Y. Sakai, T. Tominaga, A. Fukuoda, T. Kimura, and M. Ichikawa, *J. Radioanal. Nucl. Chem.*, 137 (1989) 287.
145.   H. A. Stöckler, H. Sano, and R. H. Herber, *J. Chem. Phys.*, 45 (1966) 1182.
146.   M. O'Rourke and C. Curran, *J. Am. Chem. Soc.*, 92 (1970) 1501.
147.   M. K. Friedel, B. F. Hoskins, R. L. Martin, and S. A. Mason, *J. Chem. Soc. D* (1970) 400.
148.   K. C. Molloy and K. Quill, *Polyhedron*, 5 (1986) 1771.
149.   H. A. Stöckler, H. Sano, and R. H. Herber, *J. Chem. Phys.*, 46 (1967) 2020.
150.   C. S. Frampton and J. Silver, *Inorg. Chim. Acta*, 112 (1986) 203.
151.   M. Hanack, J. Metz, and G. Pawlowski, *Chem. Ber.*, 115 (1982) 2836.
152.   A. Nath, M. Harpold, M. P. Klein, and W. Kündig, *Chem. Phys. Lett.*, 2 (1968) 471.
153.   J. L. Thompson, J. Ching, and E. Y. Fung, *Radiochim. Acta*, 18 (1972) 57.
154.   C. Ercolani, C. Neri, and P. Porta, *Inorg. Chim. Acta*, 1 (1967) 415.
155.   F. Pomposo, D. Carruthers, and D. Stynes, *Inorg. Chem.*, 21 (1982) 4245.
156.   J. L Petersen, C. S. Schramm, D. R. Stojakovic, B. M. Hoffmann, and T. J. Marks, *J. Am. Chem. Soc.*, 99 (1977) 286.
157.   J. G. Stevens and W. L. Gettys, in *Mössbauer Isomer Shifts*, G. K. Shenoy and F. E. Wagner, Eds., North-Holland, Amsterdam, 1978.

3

Synthesis and Spectroscopic Properties of Phthalocyanine Analogues

Nagao Kobayashi

A. INTRODUCTION

In this chapter, the chemistry of phthalocyanine (Pc) analogues will be described. Generally, the Pc analogues that will be treated here include porphyrin compounds that have either four nitrogens at the *meso* positions (tetraazaporphyrins), four aromatic rings connected directly to pyrrole rings in a coplanar fashion, or with both *meso*-nitrogens and aromatic rings, together with their dimers. Accordingly, porphyrins containing only one, two, or three nitrogens without fused coplanar aromatic rings are eliminated for the reason that they are closer to porphyrins rather than Pcs. Compared to Pcs, the physicochemical properties of Pc analogues are not always well elucidated; in many cases the electronic absorption spectra alone are reported. Hence, in each group of compounds, the general synthetic procedure and properties concerned with synthesis (such as solubility) are first described, followed by the physicochemical properties.

Although the structures of many of the Pc analogues so far reported are shown in this review, the list is not complete. According to a catalogue recently published by Luk'yanets [1], there are indeed several more Pc analogues, the data of which the author could not access since they have been published mostly as patents or in Russian journals. This chapter is based on literature collected by the author.

B. TETRAAZAPORPHYRINS

The first tetraazaporphyrins (TAPs) were synthesized in 1937. Linstead and Cook reacted diphenylmaleonitrile and magnesium powder at 275°C for 10 min, and obtained **1** at 92% yield [2]. Demetallation of **1** was achieved in 15% HCl, to produce **2**. Copper **3** [2], manganese **4** [3], and vanadyl **5** [4] derivatives were prepared similarly as for **1**. Since then, the

Tetraazaporphyrin   Structures

1	$R_1=R_2=$Ph-	M=Mg
2		M=H$_2$
3		M=Cu
4		M=Mn
5		M=VO
6	$R_1=R_2=$H-	M=Mg
7		M=H$_2$
8		M=Cu or Ni
9	$R_1=$CH$_3$, $R_2=$H-	M=Mg
10		M=H$_2$
11	$R_1=R_2=$CH$_3$-	M=Mg
12		M=H$_2$
13		M=Cu, Ni, or Zn
14	$(R_1, R_2)=$-(CH$_2$)$_4$-	M=Mg
15		M=H$_2$
16		M=Cu, Pd, Ni, Co
17	$R_1=$(CH$_3$)$_3$C-, $R_2=$H-	M=Mg
18		M=H$_2$
19		M=VO
20		M=Ni or Pd
21		M=Co
22		M=Cu
23	$(R_1,R_2)=$	M=H$_2$

24		M=Mg, VO, Fe, Co, Cu, or Pd
25	$R_1=R_2=$(CH$_3$)$_3$C-Ph-	M=H$_2$
26		M=Mg, VO, Fe, Ni, or Cu
27	$R_1=$(CH$_3$)$_3$CO-, $R_2=$H-	M=H$_2$
28	$R_1=$n-C$_5$H$_{11}$O-, $R_2=$H-	M=Mg
29		M=H$_2$ or Mg
30	$R_1=$n-C$_5$H$_{11}$S-, $R_2=$H-	M=Mg

31	$R_1=R_2=$n-C$_5$H$_{11}$S-	M=Mg
32	$R_1=$(CH$_3$)$_3$CNH-, $R_2=$H-	M=Mg
33	$R_1=$CH$_2$=CHCH$_2$NH- $R_2=$H-	M=Mg
34	$R_1=$(C$_2$H$_5$)$_2$N-, $R_2=$H-	M=Mg
35	$R_1=R_2=$(CH$_3$)$_3$CNH-	M=Mg
36	$R_1=$(CH$_3$)$_3$C-, $R_2=$-CO$_2$H	M=VO
37	$R_1=$(CH$_3$)$_3$C-, $R_2=$-CN	M=H$_2$
38		M=VO
39		M=Cu
40	$R_1=$Ph-, $R_2=$-CN	M=Mg
41	$R_1=$(CH$_3$)$_2$N-Ph-, $R_2=$-CN	M=Mg
42	$R_1=$(CH$_3$)$_3$CNH-, $R_2=$-CN	M=H$_2$
43		M=Mg
44		M=Cu
45	R_1(CH$_3$)$_3$C-, $R_2=$-NO$_2$, H$_3$-	M=H$_2$
46	$R_2=$-(NO$_2$)$_2$, H$_2$-	M=H$_2$
47	$R_2=$-(NO$_2$)$_3$, H-	M=H$_2$
48	$R_2=$-NO$_2$	M=H$_2$
49	$R_2=$-(NO$_2$)$_2$, H$_2$-	M=Zn
50	$R_2=$-(NO$_2$)$_3$, H-	M=Zn
51	$R_2=$-NO$_2$	M=Zn
52	$(R_1, R_2)=$-S-(CH$_2$)$_2$-S-	M=H$_2$
53		M=Zn, Ni, Cd, Pb Sn, Al, Cr, Ca, In, Ge, or Zn
54	$R_1=R_2=$CH$_3$S-	M=Cu
55		M=Mg
56		M=Ni
57		M=H$_2$
58	$R_1=R_2=$Ph-CH$_2$S-	M=Ni
59	$R_1=R_2=$NaS-	M=Ni

60

simplest TAPs **6-8** [5], and tetramethylated **9** and **10** [6], octamethylated **11-13** [7], and tetracyclohexeno TAPs **14-16** [8, 9] have been reported. With the exception of **1**, magnesium compounds were first obtained by heating substituted maleonitrile or succinonitrile with magnesium in an alcohol such as propyl, amyl, or pentyl alcohol, after which the magnesium was removed by treatment with acid. Compounds **8**, **13**, and **16** were then obtained by metal insertion reaction into metal-free **7**, **12**, and **15**, respectively. These compounds showed, however, very limited solubility in many organic solvents, which has prevented their detailed characterization by spectroscopy. In this respect, the Russian workers, especially the group of Luk'yanets, began to report many soluble TAPs from the beginning of the 1970s. Compounds **17-51** have all been synthesized by Russian researchers. *Tert*-butylated **17** was obtained in 70% yield by the reaction of *tert*-butylmaleonitrile and magnesium in propyl alcohol, and demetallation in boiling acetic acid produced **18** in 80% yield [10a]. Luk'yanets et al. recently succeeded in separating several isomers from the randomers of **18** [11]. These compounds are highly soluble in many organic solvents, and the absorption spectra are the sharpest among those TAPs so far synthesized. Tetra-2,3-(dibenzobarreleno)TAPs **23** and **24** were obtained similarly [12] using dibenzo[5,6;7,8]barrelene-2,3-dicarbonitrile that was obtained from anthracene and acetylenedicarbonitrile, and are soluble in many solvents. Compounds **26** were obtained directly from bis(*p-tert*-butylphenyl)maleonitrile and metal or metal salts, and are more soluble than **1-5** [13]. Alkylamino-, alkoxy-, and alkylthio-substituted **27-35** were prepared in 15-45% yield [14, 15]. The starting materials (i.e., alkylamino-, alkoxy-, and alkylthio-substituted maleo (or fumaro) nitriles) were obtained by the reaction of monochloro- and dichloromaleo (or fumaro) nitriles with alkylamine, sodium alkoxide, and sodium alkylsulfide. Since two cyano groups strongly withdraw electrons, such substitution reactions proceed relatively smoothly. Compounds **36-44** were obtained in 23-67% yield, and are interesting in that both electron-releasing and -withdrawing groups are attached in close proximity in a molecule [16]. The nitro group-introduced **45-51** were synthesized mainly in order to examine their fluorescence properties [17]. A TAP containing eight sulfurs, **52**, was synthesized for the first time in 1960 [18], and recently various metal-inserted tetra-(1,4-dithiacyclohexeno)TAPs **53** have been reported [19]. In order to synthesize precursors for octaalkylthio-substituted TAPs, not only **52** and **53**, but also **54-57** [20] and **58** [21], disodium and dipotassium salts of 1,2-dimercapto-1,2-dicyanoethylene are often utilized. These dinitriles react with alkyl halides to produce octaalkylthio-substituted malononitriles. The latter were reacted directly with metals or metal salts to produce **53-56** [19], or further changed to diimino derivatives with ammonia gas to yield nonmetallated TAPs such as **52** [18]. Metal-free **57** was obtained by removal of magnesium from **55** [20]. A starlike TAP with four tin atoms **60** was recently prepared as deep green-black needles from **58** via **59** in 55% yield, and its structure was determined by x-ray crystallography [21]. Each tin atom is coordinated to two thiolates of different pyrrole rings, and two tins lie 0.73 or 0.54 Å above the TAP-thiolate plane, while other two tins lie below this plane, resulting in a quasichair conformation.

NMR data for TAP have been reported only for **18** [11]. β-Protons of pyrrole, protons of the *tert*-butyl group, and NH protons appear at ~8.9, 2.2, and -2.4~-2.5 ppm, respectively. In the case of 4,4',4",4"'-*tert*-butylated H_2Pc [4], protons of the *tert*-butyl group and NH emerge at ~1.9 and -2.1~-2.2 ppm, respectively. Accordingly, the values for NH protons suggest that the ring current effect of TAP is larger than with Pc.

IR data for TAPs are very scarce. Some frequency numbers have been reported in [22], but no spectra are seen in the literature. Figure 1 shows IR data of the tetra-*tert*-butylated TAPs **17-19, 21,** and **22,** which we have recently synthesized [4]. Of these, the bands at 2956-2960 ($v_{as}CH$), 2904-2910, 2864-2870 (v_sCH), 1247-1260 (skeleton), and 1200-1208 cm^{-1} are attributed to *tert*-butyl groups, but other bands observed commonly for MtTAPs at 772-777, 843-850, 999-1013, 1064-1078, 1360-1362, 1390-1392, and 1459-1483 cm^{-1} would be mostly due to skeletal vibrations of MtTAPs. Among these, the bands at 999-1013 cm^{-1} are the strongest, and the bands at 1064-1078 and 1459-1483 cm^{-1} are doublets. Nonmetallated **18** displays, in addition to a band characteristic of NH vibration at 3298 cm^{-1}, bands at 736 and 940 cm^{-1} that are not seen with metallated derivatives.

Electronic absorption and magnetic circular dichroism (MCD) spectra of **18, 19,** and **22** are shown in Fig. 2. The so-called *Q* band of **18** consists of two strong bands at 619 and 551 nm, together with several vibrational bands. When metallated, a single band appears at around the center of these two bands [10a]. Electronic absorption data accumulated so far indicate that the absorption coefficients (ε) of TAPs are generally smaller than those of Pcs having the same central metal and peripheral substituent groups [1-23]. MCD spectra in the *Q* region can be easily rationalized as with Pcs [23] and porphyrins [24]. Compound **18** shows an MCD trough and peak associated with the above two absorption bands, which are attributed to Faraday *B* terms, since there are no degenerate excited states [25]. In contrast, *S*-shaped *Q*-band MCD waves observed for metallated compounds are mainly Faraday *A* terms in origin, since the excited state is doubly degenerate under the D_{4h} approximation [23]. Compounds **18, 19,** and **22** show another strong absorption band at 300-350 nm. Although these bands had generally been thought to be the Soret band, this is in fact not the case, as is substantiated in Fig. 3.

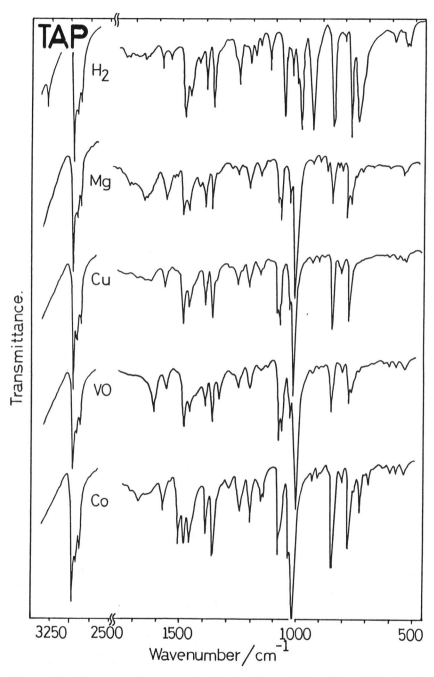

Figure 1 IR spectra of tetra-*tert*-butylated TAPs (**17-19, 21**, and **22**) in KBr disk.

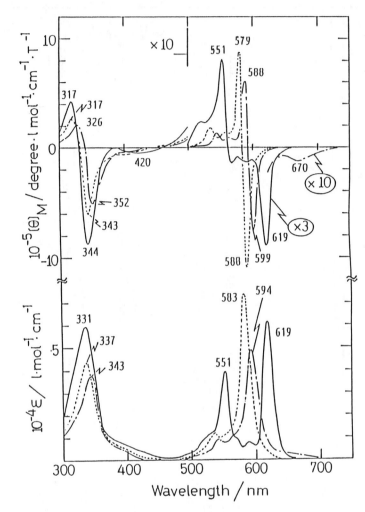

Figure 2   Absorption (bottom) and MCD (top) spectra of **18** (solid lines), **19** (dotted broken lines), and **22** (dotted lines) in pyridine.

Figure 3 shows the absorption and MCD spectra of **1** and **17**. It is recognized first that the intensity and positions of the bands differ greatly between the two compounds. The bands of **1** are much more intense than those of **17,** and are located at longer wavelengths. Second, **1**, which has more perfect D_{4h} symmetry than **17**, exhibits a very clear Faraday A term associated with the absorption peak at 460 nm. Accordingly, if we accept Gouterman's "four orbital theory" [26], this is the B (Soret) band. The

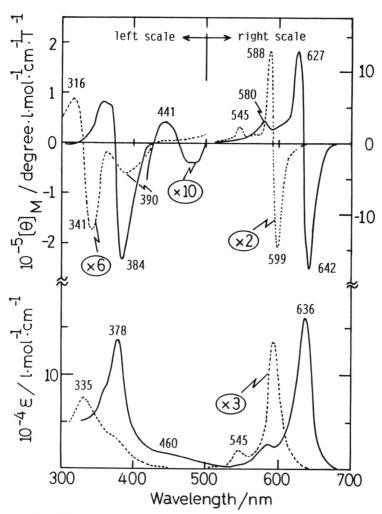

Figure 3 Absorption (bottom) and MCD (top) spectra of **1** (solid lines) and **17** (dotted lines) in pyridine. Note the encircled magnification factor.

weak absorption shoulders seen at around 400 nm for **18, 19,** and **22** (Fig. 2) appear to correspond to the Soret bands. Since the symmetry of **18, 19,** and **22** is a less than perfect *D4h* compared with **1,** and these compounds are mixtures of four positional isomers, no explicit Faraday *A* term appears to be observed. Compared with Pcs, the position and intensity of the absorption spectra of TAPs seem to be more easily altered by solvents [27] and substituent groups [1-23]. For example, the *Q*-band ε value of **17** changes from 48,000 in pyridine [4] to 123,000 in hexane [4, 10a], and the most intense Q bands of **40-44** lie at around 800 nm [16]. If we compare their Q-band positions with those of the compounds in Fig. 2, for example, the difference is more than 200 nm. Possibly, large electronic transition moments might be induced by the attachment of both electron-withdrawing and -releasing groups, as seen in substituted benzenes [28]. The oxygenated complex of **4** is distinct from other MtTAPs in that it shows the Q band at an unusually long wavelength (*e.g.*, at 725 nm in DMF [3]).

The absorption spectra of **18** are reported also in a polymer solution and Langmuir film [29]. By detecting a Stark effect (relative change in transmittance), Blinov et al. claimed that a new band that appeared at 600-650 nm in the Langmuir film arose from a dense packing of **18**, and was due to intermolecular charge transfer. By the introduction of sulfur as peripheral substituent groups, the absorption spectra broaden and shift to longer wavelengths [20, 21] This observation is well accepted by comparing the spectra in Figs. 2 and 3 with the spectrum of **58** [21]. The main Q band of **58** lies around 650 nm. Interestingly, coordination of the dialkyltin group to the *meso*-nitrogens of **58** doubles the absorbance. This phenomenon has been attributed to the stabilization of the second highest molecular orbital of TAP, which has a high orbital density on the *meso*-nitrogens. One of the broad absorption peaks between the Soret and Q bands observed commonly among compounds **53** has been attributed to an n-π^* transition (*n:* nonbonding electron of sulfur) [19]. Dodd and Hush reported the absorption spectra of mono- and dinegative species of **1** [30a]. The color of the parent **1** is emerald, but that of the mononegative ion is grey, and it exhibits strong absorptions at ~700 and 440 nm, together with two weak bands at ~850 and 970 nm. The dinegative species shows the first and second strongest bands at ~280 and 1000 nm, respectively, and is brown in color. The absorption spectrum of the mononegative ion resembles in shape that of CuPc shown in the same paper, but that of the dinegative species is very different from the dinegative CuPc species, although the reduction appears to be occurring at the macrocycles. They conclude that, on conversion of the mononegative ions to dinegative ions, the absorption bands of TAPs are shifted to higher frequencies, but that those of Pcs are shifted to lower frequencies. However, the assignment of the bands of these reduced species was not attempted. Hydrogenation reactions of some H_2TAP and MgTAP species were performed with a palladium black catalyst, and the spectra of the resultant species are shown [31]. Tetrahydro-**2** and -**12** exhibit two intense absorption bands at 500-540 and 670-700 nm in chlorobenzene. Similar absorption spectra have been reported for D_{2h}-type tetrahydroporphyrins [32, 33].

Theoretical molecular-orbital calculations of TAPs do, of course, exist [27, 34-38]. The energy difference of the two top-filled orbitals of

MtTAPs, a_{1u} and a_{2u} orbitals, using the D_{4h} approximation, are larger than for common porphyrins, but smaller than with Pcs. The Q band energy is higher, and the calculated oscillator strengths at both the Soret and Q bands are generally smaller than in Pcs, as has been observed. In addition, there is a fairly large number of n-π^* transitions predicted to be in the region of the Q and Soret bands. In [27], the intense absorption of MgTAP at around 330 nm was assigned to the N band, and its long-wavelength shoulder to the B (Soret) band. For MtTAPs, the calculations predict the intensification of the configuration mixing of the B and N states, which results in weakening of the B bands and strengthening of the N bands [27]. This is consistent with our data in Figs. 2 and 3. Ellis et al. [38] reported a charge distribution map for H$_2$TAP and spin-density maps for CuTAP and FeTAP. Although the twofold $e_{(xz, yz)}$ level of D_{4h} symmetry splits into b_{2g}, b_{3g} levels in D_{2h}, they assign $b_{1u} \rightarrow b_{2g}, b_{3g}$ and $a_u \rightarrow b_{2g}, b_{3g}$ transitions to Q_x and Q_y bands, respectively, of H$_2$TAP. That is, they placed the b_{1u} orbital as the highest occupied molecular orbital (HOMO), in marked contrast to other papers [27, 31, 34-36] where the a_u orbital is assumed to be the HOMO. The calculated band position of the Soret (446-498 nm), Q_x (794 nm), and Q_y (595 nm) bands are obviously somewhat lower compared with experimental data [1-15, 17-22]. They assumed the presence of a π-d forbidden transition in the longer-wavelength side of the Q band of CuTAP.

The luminescent properties of **6** [39], **2** and its zinc complex [36], **18** and its zinc complex, and **45-51** [17] have been examined by Russian workers. All of these complexes fluoresce from the S_1 level and phosphoresce from the T_1 level at 77 K, and the fluorescence spectra of **2** and **18** are mirror images of the lowest Q band. Introduction of nitro groups at the β positions of the TAP skeleton does not appreciably influence the photophysics (quantum yield ϕ_F, lifetime τ, and so on) of the molecule. Compared with the ϕ_F (=0.66 in CHCl$_3$) and τ (=7.02 ns) values of *tert*-butylated H$_2$Pc [4], the values for **18** (ϕ_F=0.18, τ=3.0 ns) [17] are much smaller. They pointed out, from the change of fluorescence polarization, that there are hidden n-π^* transitions in the 300-400 nm region of the ZnTAP spectrum. Quite recently, we have found that **1** emits from both the Soret (S_2) and Q (S_1) states [4]. The S_2 emission peak lies at ~470 nm.

ESR spectra of **54** in concentrated sulfuric acid and 1-chloronaphthalene at 77 K have been reported [20]. At 1 mM concentration, **54** exists as a monomer in the former solvent, but mostly as a cofacial dimer in the latter solvent. Since ESR data on MtTAPs have not been reported in the literature, we synthesized and recorded the ESR spectra of **19, 21**, and **22** [4]. Spectra of **19** and **21** are shown in Fig. 4, and the obtained ESR parameters are collected in Table 1, together with those of *tert*-butylated

Table 1. ESR Parameters of Co, VO, and Cu Complexes of Tetra-*tert*-Butylated Tetraazaporphyrin (BTAP), Phthalocyanine (BPc), and Naphthalocyanine (BNc) in Toluene at 77K

compound	$g_{//}$	$g_{\perp}$	$10^4 A^{Mt}_{//}$	$10^4 A^{Mt}_{\perp}$	$10^4 A^{N}_{//}$	$10^4 A^{N}_{\perp}$
				(cm^{-1})		
CoBTAP	2.0259	2.8659	167.63[a]	235.81[a]		
			min 117.99	min 192.00[b]		
			max 215.31	max 267.86		
CoBPc	2.0244	2.8051	156.69[a]	217.94[a]		
			min 116.14	min 186.00[b]		
			max 197.96	max 251.06		
CoBNc[c]						
CoBAc[c]						
CoBTAP(py)	1.9980	2.26[b]	99.31		1 7.10	
CoBPc(py)	2.0036	2.34[b]	93.88		1 7.19	
CoBNc(py)	2.0067	2.26[b]	91.24		1 7.03	
VOBTAP	1.9659	1.9889	157.33[a]	53.80[a]		
			min 155.03	min 38.43		
			max 160.28	max 67.02		
VOBPc	1.9652	1.9876	157.60[a]	55.08[a]		
			min 154.33	min 40.61		
			max 159.10	max 69.82		
VOBNc	1.9667	1.9878	158.74[a]	54.55[a]		
			min 156.78	min 41.03		
			max 161.35	max 68.18		
CuBTAP	2.1511	2.0472	218.33		15.58	17.11
CuBPc	2.1582	2.0479	214.89		15.03	16.43
CuBNc	2.1621	2.0485	213.09		15.42	15.73

[a] Average value, [b] Approximate value, [c] Parameters could not be obtained because of aggregation.

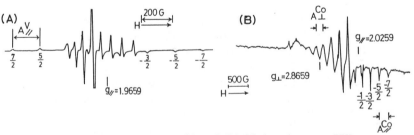

Figure 4   ESR spectra of (A) **19** and (B) **21** in toluene at 77K.

Pcs and naphthalocyanines (Ncs). The ESR parameters are not very different from those of Pcs and Ncs.

Magnetic susceptibility measurements from 5 to 300 K were performed for **4** [3]. Although excellent Curie-Weiss behavior was found at temperatures lower than 45 K, the spin state of manganese in this complex could not be determined. Manganese in Mn(II)Pc has an unusual $S=3/2$ spin state [40], while that in Mn(II) tetraphenylporphyrin has an $S=5/2$ ground state [41].

Berezin et al. [42] reported that the stability of the tetraazaporphyrin macrocycle in sulfuric acid is significantly lower than that of the Pc ring.

Compounds **53** have semiconductor properties: The specific electrical conductivities range from 10^{-11} to 10^{-9} S/cm [19]. Four reduction potentials of **1** have been determined in DMF [30b,c]. Of the eight kinds of porphyrins examined, this compound is the easiest to reduce; in other words, each reduction lies at a more anodic potential than those of other porphyrins. Quite recently, the mesogenic properties of octa(octylthio)-TAPs were reported [43]. The, Zn, Cu, Ni, and Co complexes form a discotic mesophase, but the columnar hexagonal arrangement was a function of the central metal, being disordered except for the Co complex. The μ-oxo complex of *tert*-butyl FeTAP and **21** were found to be more effective than various Co- and FePcs as catalysts for mild ligand phase oxidation of cyclohexane by cumene hydroxide [10b].

C. NAPHTHALOCYANINES

Compounds **61-68** are conventionally called 1,2-naphthalocyanines (1,2-Ncs), since they are synthesized from 1,2-dicyanonaphthalene or 1,2-naphthalic anhydride. Although the structure of one isomer is shown in a scheme, these are, at least statistically, mixtures of four positional isomers. Linstead et al. heated 1,2-dicyanonaphthalene and etched magnesium turnings at 370°C for 80 min, and obtained **61** as a mixture of a bright green, ether-soluble α form and a dark green insoluble β form in the ratio of 2.5:1 [44]. The α form of **61** is soluble in ethyl alcohol, acetone, ethyl acetate, and chloroform, and is extremely soluble in ether. Treatment of the α or β form of **61** with concentrated sulfuric acid gives the α or β form of **62**, respectively. The α form of **62** is soluble in boiling aniline but not in low-boiling solvents, and the β form was purified by crystallization from chloronaphthalene. Copper salts and zinc dust react with 1,2-dicyanonaphthalene at 270-300°C and, after crystallization from chloronaphthalene, give **63** and **64**. These two compounds can be obtained also by the metal exchange reaction of **61**. The 1,2-Ncs show an even greater stability to heat and reagents than the Pcs. In particular, they are only

Naphthalocyanine Structures

6 1	M=Mg
6 2	M=H$_2$
6 3	M=Cu
6 4	M=Zn
6 5	M=VO, AlCl, Ga
6 6	M=Fe[(CH$_3$)$_3$C-NC-]$_2$

6 7 M=Fe

6 8 M=Fe

6 9	R$_1$-R$_6$=H-	M=H$_2$
7 0		M=Mg

7 1		M=Zn
7 2		M=VO, Co, Ni
7 3		M=Cu, Pb, Pd
7 4		M=AlOH
7 5		M=GaOH
7 6		M=SnCl$_2$
7 7		M=Mn(OCOCH$_3$)
7 8		M=Fe
7 9		M=Fe[(CH$_3$)$_3$C-NC-]$_2$

8 0 M=Fe

8 1 M=Fe

8 2 M=Fe

8 3 M=Fe

8 4	R$_3$=(CH$_3$)$_3$C-	M=H$_2$
	other Rs=H-	
8 5		M=VO
8 6		M=Zn
8 7		M=Cu
8 8		M=Co
8 9		M=Ni, Pb
9 0		M=SnCl$_2$
9 1		M=Mn(OCOCH$_3$)
9 2	R$_3$=1-Adamantyl	M=VO or Cu
	other Rs=H-	
9 3	(R$_3$, R$_4$)=	M=H$_2$

9 4		M=Mg, VO, Co, Cu, or AlCl
9 5	R$_1$=Ph,	M=H$_2$
	other Rs=H-	
9 6		M=VO, Mn, Cu or ?
9 7	R$_1$=(CH$_3$)$_3$C-Ph-	M=VO or Zn
	R$_2$=(CH$_3$)$_3$C-	
9 8	R$_3$=R$_4$=Cl-,	M=VO or Cu
	other Rs=H-	
9 9	R$_3$=R$_4$=Br	M=VO or Cu
1 0 0	R$_1$=R$_6$=OR	M=H$_2$
	other Rs=H-	
1 0 1		M=Cu
1 0 2	R$_1$-R$_6$=H	M=Ge(OSi(C$_2$H$_5$)$_3$)$_2$
1 0 3		M=Ge(OSi(n-C$_4$H$_9$)$_3$)$_2$
1 0 4		M=Si(OSi(n-C$_6$H$_{13}$)$_3$)$_2$
1 0 5		M=Si(OSi(n-C$_4$H$_9$)$_3$)$_2$

slowly oxidized by nitric acid and ceric sulfate. The synthetic procedure of **65** could not be obtained, but appears to be essentially similar to that for **61**, **63**, and **64** [45]. The iron complex of 1,2-Nc (i.e., Fe-1,2-Nc), a parent complex of **66-68**, was prepared in ~30% yield by reaction of 1,2-naphthalenedinitrile with pentacarbonyliron in refluxing ethylene glycol or 1-chloronaphthalene under nitrogen for ~1 h [46], as reported for the preparation of FeFcs [47]. Compounds **66-68** were then obtained by reacting Fe-1,2-Nc with excess *tert*-butyl-, cyclohexyl-, or benzylisocyanide for many hours at 50-60°C, followed by purification using silica-gel chromatography. The x-ray structure of one of the isomers of **67** is shown in [46], where four naphthalene units are attached to a TAP skeleton in a clockwise direction.

There are few data available on spectroscopic properties of 1,2-Ncs. Well-resolved ^{1}H NMR spectra are shown for **66-68** [46]. The signals from aromatic protons appear in the region of the so-called aromatic protons, suggesting that the iron in **66-68** is in a divalent low-spin state. Yurlova [48] reported the IR spectrum of **63** and its properties as an organic semiconductor with an activation energy of 0.89 eV. This value is just inbetween that of CuPc (1.2 eV) and Cu-1,2-Ac, **106**, (0.72 eV) [48]. The positions of the Soret and *Q* bands are reported for **66-68** and compared with those of common FePcs. Although the Soret bands of **66-68** are shifted to longer wavelengths by ~20-30 nm, the *Q* band positions are roughly the same. Not only these data but also other data accumulated to date indicate that the *Q* band does not shift markedly unless the aromatic rings are fused to the TAP skeleton radially, as seen from the center. Two Mössbauer parameters (the isomer shift and the quadrupole splitting) are collected for Fe-1,2-Nc [46]. They are smaller than those of FePc [49].

Compounds **69-105** are generally called naphthalocyanines, or when necessary to discriminate from 1,2-Ncs, they are called 2,3-Ncs. The most simple non-metallated 2,3-Nc, **69**, was obtained by boiling 2,3-dicyanonaphthalene and sodium isoamylate in amylalcohol, and subsequent replacement of two sodiums by two protons, and is insoluble in the majority of organic solvents [50]. Compounds **70-78** were provided by fusing the nitrile and the appropriate metal salts or metal dust with or without solvent (bromonaphthalene). If required, ammonium molybdate was used as a catalyst [50]. Vogler described another method for preparing **71** [51]: naphthalene-2,3-dicarboxylic acid, zinc salt, and excess urea were heated at 310°C for 40 min to give **71** in 18% yield. Pentacarbonyliron and 2,3-dicyanonaphthalene also give **78** in ~20% yield when boiled in ethylene glycol [49, 52]. Hanack et al. synthesized **79-83** by reaction of **78** with excess *tert*-butylisocyanate, cyclohexylisocyanate, benzylisocyanate, 4,4'-bipyridyl, or pyridine [53]. *Tert*-butylated **84-91** were first prepared in yields ranging from 20 to 60% in 1971 [54]. The starting material, 6-*tert*-

butyl-2,3-dicyanonaphthalene, was prepared from 4-*tert*-butyl-*o*-xylene [55] via 1,2-bis(dibromomethyl)-4-*tert*-butylbenzene in the usual manner [56]. As in the case of **69, 84** was obtained by heating the nitrile with sodium isoamylate in isoamylalcohol, but **85-91** were obtained by fusing the nitrile in the presence of the appropriate metal dust or metal salts, occasionally with ammonium molybdate and in solvents. Owing to the *tert*-butyl group, these compounds are soluble in many solvents and accordingly could be purified by the use of columns. The manganese complex is often obtained with an axial ligand, as shown in **91**. In this case, manganese may be in the trivalent state. Although not listed, Gal'pern et al. [57] reported Ncs that have both *tert*-butyl groups and bromines as R3 and R5, respectively. They say that introduction of bromine heightens the relative stability in solution. Compound **92**, having adamantyl substituents, were synthesized essentially similarly to **84, 85,** and **87** [58]. Tetra-6,7-dibenzobarreleno-2,3-Ncs **93** and **94** were reported in 1987 [59]. The corresponding Pc [60] and TAP [12] are also known. Two methods were developed for the synthesis of the starting dinitrile, 6,7-dibenzobarreleno-2,3-naphthalenedicarbonitrile. However, the method starting from 2,3-dimethyltryptycene is superior. Bromination of 2,3-dimethyltryptycene to its tetrabromide derivative, and the following reaction with fumaronitrile produced the above dinitrile in 76% total yield. Metallated **94** were prepared in 1-bromonaphthalene by heating the nitrile with metal salts in the presence of urea, sodium sulfate, and catalytic amounts of ammonium molybdate in yields from 10 to 50%. Metal-free **93** was obtained by treatment of the magnesium complex with acid. Ncs with phenyl groups as substituents, **95-97**, appeared in [61]. The starting material for **95** and **96**, 1-phenyl-2,3-naphthalenedicarboxylic anhydride, was obtained very easily by heating commercially available phenylpropiolic acid in acetic anhydride [62]. The other starting material, 1-phenyl-2,3-naphthalenedicarbonitrile, could be obtained via the anhydride by converting it to the corresponding imide, and amide, and subsequent dehydration of the latter. Compounds **96** were obtained by treatment of either the anhydride with metal salts in the presence of excess urea (plus solvents) or the nitrile with metal salts in 30-70% yield. Metal-free **95** was prepared from the zinc complex by demetallation with acid. The solubilities of **95** are somewhat higher than the *tert*-butylated Ncs **85-91**, and those of **97** are even ~100 times higher than **95**. Halogen-atom-containing **98** and **99** were also prepared [63, 64]. These compounds were synthesized in 30-50% yield by fusing halogenated 2,3-dicyanonaphthalene in the presence of metal salts and occasionally ammonium molybdate and a solvent (1-chloronaphthalene). Reference [63] also introduces the preparation of tetra- and heptadecachlorinated CuNcs and VONcs having four nitro, amino, or acetamide groups (the degree of chlorination was estimated by elemental analysis, although it is not easy to deduce the position of the 17th chlorine). Heptadeca-chlorinated CuNc alone was obtained by chlorination of CuNc by thionyl chloride using aluminium chloride as a catalyst. Octa(*n*-alkoxy) Ncs **100** and **101** were reported recently [65]. Ethyl, propyl, butyl, pentyl, and octyl chains were introduced as R. Pcs with alkoxy groups at the same positions are also known [65, 66]. The starting materials, 1,4-dialkoxy-2,3-dicyanonaphthalenes, were prepared in 40-80% yield from 2,3-dicyanonaphthalene-1,4-diol and alkyl iodide in the presence of potassium carbonate. Metal-free **100** was prepared by utilizing lithium metal in pentanol or butyl-lithium in tetrahydrofuran in yields ranging from 10 to 30%. Compounds **102-105** are characteristic in that they have covalently bound very bulky trialkylsiloxy groups [67, 68]. The start of the synthetic pathway is the reaction of 1,3-diiminobenz(*f*)isoindoline with germanium, or silicon tetrachloride in a mixture of tri-*n*-butylamine and tetrahydronaphthalene. The resultant Ncs containing two chlorines as axial ligands were then hydrolyzed to give Ncs with two hydroxo groups. These were subsequently reacted with trialkylchlorosilane, often in the presence of a base such as tri-*n*-butylamine.

The ^{1}H NMR spectrum of ZnNc, **71**, was recently recorded in deuterated DMSO [69]. The 1,4, 5,8, and 6,7 protons appeared at 7.17 (singlet), 6.14 (quartet), and 5.38 (quartet) ppm, respectively. In [68], the spectrum of **104** in CDCl$_3$ is shown. In this case, the 1,4, 5,8, and 6,7 protons appear at ~9.96, 8.54, and 7.77 ppm, respectively. Protons from alkyl groups appeared between ~-2.5 and 0.5 ppm because of the ring current effect of the Nc ring. Thus the positions of aromatic protons shift significantly with the kind of central metal. We have recorded the 500 MHz ^{1}H NMR spectrum of **84** in deuterated pyridine [4]. 1,4 Protons, protons of the *tert*-butyl group, and NH protons give signals at ~8.87, 1.78, and -1.64 ppm, respectively. Comparing the positions of NH protons among *tert*-butylated TAP (-2.47), Pc (-2.17), and Nc, it is noticed that the ring current decreases with the enlargement of the size of the π systems.

The IR spectrum of **63** is shown in [48], and the spectra of a few MtNcs are seen in [70], but are not well explained. We have, therefore, measured IR spectra of several *tert*-butylated Ncs (Fig. 5) [4]. They are, at a glance, rather simpler than those of the TAPs (Fig. 1) and Pcs (Fig. 6). Except for the bands assignable to *tert*-butyl groups, the characteristic bands common to MtNcs are observed at 470-472, 724-725, 742-749, 808-812, 888-901, 946-947, 1082-1088, 1100-1104, 1142-1144, and 1343-1359 cm^{-1}. Of these, the bands at 888-901 cm^{-1} are a doublet, and those at 1343-1359 cm^{-1} are a multiplet. The bands at 1082-1088, 1100-1104, and 1343-1359 cm^{-1} are particularly intense. The bands at 470-472 cm^{-1} may be attributed to skeletal vibrations of naphthalene [71]. IR spectra of **104** deposited on KBr or thalium substrates are shown in [72]. Three peaks

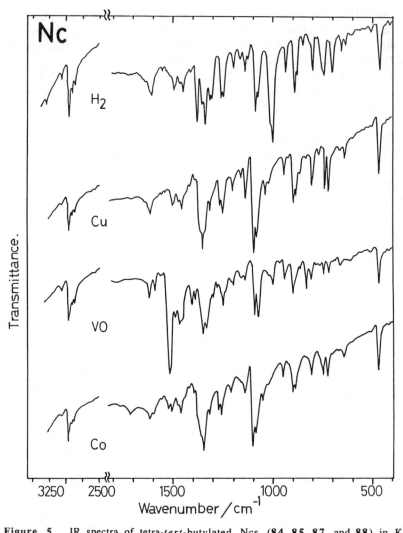

Figure 5   IR spectra of tetra-*tert*-butylated Ncs (**84, 85, 87**, and **88**) in KBr disk.

were recognized between 800 and 1050 cm^{-1}, and they were assigned to Si-C stretching, CH_3 rocking, and Si-O stretching modes, while two peaks at 700-800 cm^{-1} were attributed to a C-H out-of-plane bending mode.

Electronic absorption spectra of 2,3-Ncs have been reported in a number of publications [50-54, 57-59, 61, 63-65, 67, 68, 72-80]. The *Q* band absorption occurs at about 80-100 nm to longer wavelengths than

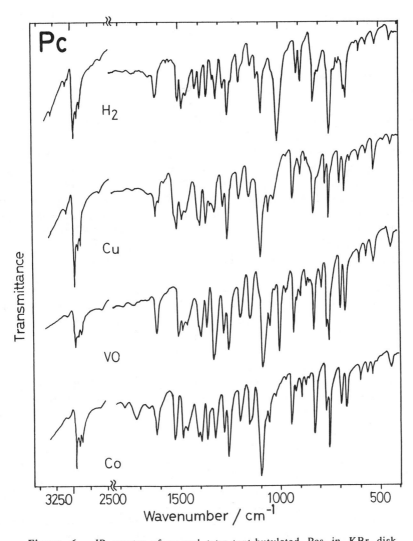

Figure 6 IR spectra of several tetra-*tert*-butylated Pcs in KBr disk.

those of similar compounds in the Pc series [23] (i.e., between 750 and 840 nm). This observation is to be expected due to the larger aromatic system of the Ncs. The Q band maxima of most of MtNcs without peripheral substituent groups lie at wavelengths shorter than 800 nm, while those of the poly-substituted Ncs easily shift to wavelengths longer than 800 nm. The effect of substituents on the Q band wavelength is larger the closer the substituent groups are to the TAP skeleton. Thus, most of the Q oand maxima of **96, 100**, and **101** occur above 800 nm. Of particular interest in this respect is the distinct effect of alkoxy [65], amino, or acylamido groups [63]. For example, a tetraamino-substituted VONc shows the Q band maximum at 870 nm in quinoline [63], an extraordinarily long wavelength for Ncs. Also, the accumulated data [4] indicate that the absorption coefficients of the Q band of MtNcs are generally larger than those of Pcs with similar substituents. Together with the longer-wavelength shift of the Q band, the so-called Soret band region also spreads towards longer wavelengths on going from Pc to Nc, and therefore bands hidden in Pcs may appear in the Nc series. An example is seen in the absorption spectrum of a ZnNc, **71** [51]. At least two bands are detected in the longer-wavelength side of the so-called Soret band at 332 nm. The spectra of nonmetallated Ncs are different from those of metal-free TAPs and Pcs, in that they do not show four Q band peaks (Fig. 7). The absorption spectra of Nc films are also shown in several reports [67, 70, 78]. The Q band maxima shift to longer wavelengths by ~50-70 nm compared with those in solution.

Molecular orbital calculations of 1,2-Ncs and 2,3-Ncs using the valence effective Hamiltonian quantum-technique have been submitted recently [80b]. In accord with experimental data, the predicted HOMO-LUMO energy gap of 2,3-H_2Nc (1.64 eV, 757 nm) is smaller than that of 1,2-H_2Nc (1.84 eV, 675 nm).

In a strong acid such as trifluoroacetic acid or sulfuric acid, or in the presence of Lewis acids such as aluminum chloride or bromide, the Q bands move to longer wavelengths stepwise [64, 74, 80], and in an extreme case [64] the shift amounts to 480 nm. By analogy with the case of Pcs [81, 82], this shift is believed to be caused by the stepwise addition of protons, or aluminum chloride or bromide to the *meso*-nitrogens.

No MCD data on Ncs have been reported to date. We have recently measured the MCD spectra of **84, 85, 87**, and **88** [4]. All compounds show Faraday A-term-type dispersion curves associated with the main Q bands, since even nonmetallated **84** exhibits MtNc-type unsplit Q bands, as mentioned (Fig. 8). Faraday A terms corresponding to those observed around 400 nm for Pcs (B_1 band) [23, 83] were detected at 515 and 504 nm for **85** and **87**, respectively.

Luminescence data of Ncs are reported in several publications [65, 68, 77, 79, 84, 85]. Tetra-*tert*-butylated AlNc exhibits fluorescence bands between 770 and 1000 nm, with ϕ_F=0.18 and τ=2.7 ns in DMSO [76] or with τ=2.9 ns in toluene [75]. Both ϕ_F and τ values are much smaller than those of the *tert*-butylated AlPc [76]. Fluorescence and phosphorescence spectra of **104** are shown in [84]. They appear at around 750-1050 and 1300-1350 nm, respectively. Energy transfer from the triplet state of **104**

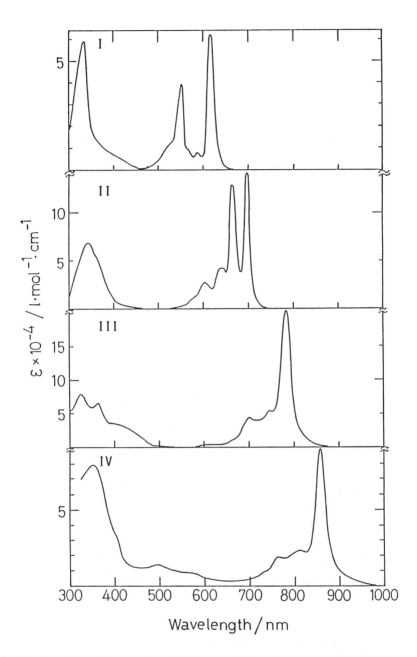

Figure 7 Absorption spectra of tetra-*tert*-butylated, non-metallated (I) tetraazaporphyrin **18**, (II) phthalocyanine, (III) naphthalocyanine **84**, and (IV) anthracyanine **112** in pyridine.

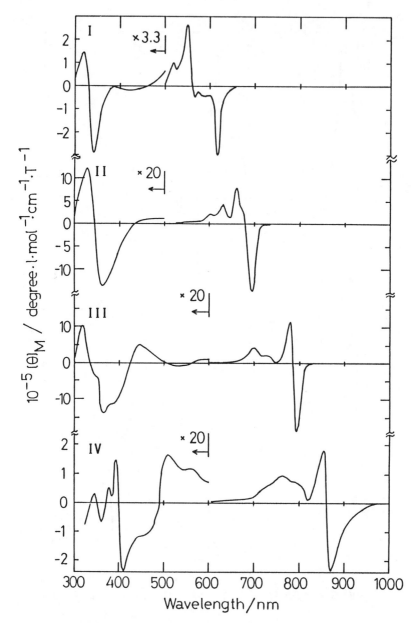

Figure 8   MCD spectra of tetra-*tert*-butylated,   non-metallated   (I)
tetraazaporphyrin **18**, (II) phthalocyanine, (III) naphthalocyanine **84**, and
(IV) anthracyanine **112** in pyridine.

to ground-state oxygen to produce singlet oxygen is studied by laser flash photolysis. This compound also produces electrogenerated chemiluminescence with a maximum at 792 nm [68]. The quantum yield of fluorescence (=0.45) and phosphorescence (=1.5 x 10^{-5}) of **105** in 2-methyltetrahydrofuran are reported in [79].

The electrical resistivities of H_2-, Pb-, and VONcs at room temperature are tabulated in [70]. They are semiconductors, and doping with bromine vapor enhances the conductivity only slightly. Mokshin et al. [86] reported that the dark conductivity of the sublimed Nc films are determined to a considerable extent by the strength of intermolecular interaction, and that the electrical conductivity of Ncs significantly exceeds that of Pcs. FeNc **78** adsorbed at carbon electrodes served as a catalyst for the electroreduction of oxygen to hydrogen peroxide or water, and the participation of an Fe(II/I) couple rather than an Fe(III/II) couple was suggested [87, 88a]. ZnNcs **71** and **86** and VONcs have been used for the light-induced oxygen reduction [88b].

The structure and molecular arrangement of VONc vacuum evaporated onto a KCl(001) surface were determined by high-resolution electron microscopy [78]. The planar VONc molecules tilt and stack in parallel along the *b* axis, which stood perpendicular to the substrate surface. In the case of VOPc, another arrangement is proposed (for details, see [78]).

Some photophysical, photochemical, and cytotoxic properties of aluminum (III) and zinc sulfonated Ncs have been reported in several publications, mainly in relation to their application to cancer therapy [89-91]. Tri- and di-sulfonated Ncs have been examined since their solubility in water is an important prerequisite.

D. ANTHRACYANINES

Compound **106** is the only 1,2-anthracyanine (1,2-Ac) so far known [48]. However, the synthetic details were not described. IR absorptions are detected at 730, 830, 890, 950, 1230, 1330, 1410, 1610, and 1670 cm^{-1}.

Most simple metallated 2,3-Acs **107** were reported in 1986 [73]. They were prepared in high (40-84%) yield by heating anthracene-2,3-dicarboxylic acid or its anhydride, metal salts, ammonium molybdate, and excess urea, occasionally in 1-chloronaphthalene at 230-260°C for several hours. Similarly, compounds **108** were produced in 43-86% yield using 9,10-dibromoanthracene-2,3-dicarboxylic anhydride [73]. The octaphenylated **109-111** were reported as the first Acs in 1971 [92], and were prepared by the nitrile method, utilizing 9,10-diphenyl-2,3-dicyanoanthracene. This starting material can be prepared by Diels-Alder addition of dicyanoacetylene to 9,10-diphenylanthracene in 82% total yield [93].

Anthracyanine Structures

107	$R_1=R_2=H-$	M=VO, Cu, or AlCl
108	$R_1=Br-, R_2=H-$	M=VO, Cu, AlCl
109	$R_1=Ph-, R_2=H-$	M=VO
110		M=Cu
111		M=AlCl
112	$R_1=H-, R_2=(CH_3)_3C-$	$M=H_2$
113		M=Co

However, since the mass production of dicyanoacetylene is difficult, this compound may not be obtained in large amounts. Also, octa-*tert*-butylated vanadium and copper derivatives were obtained from 9,10-(*p-tert*-butyl-phenyl)-2,3-dicyanoanthracene, vanadium trichloride or copper acetate, urea, and a catalytic amount of ammonium molybdate by treating at 250-275°C for ~1 h [94]. The H_2Ac 112 was obtained as a dark brown solid in 12% yield quite recently [4] by the so-called "isoindoline" method [95]. Thus, 6-*tert*-butyl-2,3-dicyanoanthracene was first reacted with ammonia gas in the presence of sodium methoxide, and then refluxed in N,N-di-methylaminoethanol for 5 h, and purified by chromatography. This H_2Ac is unstable and decomposes within several hours in solution and several days even in solid state. The cobalt complex 113 was synthesized by reaction of 6-*tert*-butyl-2,3-dicyanoanthracene and cobalt chloride in refluxing ethylene glycol for 5 h [4], also as a dark brown powder.

No NMR and IR data of Acs have been reported.

The Q-band absorption maxima of 2,3-Acs so far known occur between 830 and 935 nm [4, 64, 73, 74, 92, 94], and even metal-free 112 does not show D_{2h}-type absorption (Fig. 7) and MCD (Fig. 8) spectra [4], as in the case of *tert*-butylated H_2Nc **84**. By dissolving in sulfuric acid or in the presence of a Lewis acid such as aluminum tribromide, the Q-band peaks shift to longer wavelengths by ~310-370 nm [64], possibly by the protonation of or coordination to *meso*-nitrogens. The magnitude of the shift is generally larger than in Pcs but smaller than in Ncs [64].

Some fluorescence data of **111** are collected in [76]. In DMSO, fluorescence spectra were recorded between 900 and 1200 nm with $\phi_F < 0.1$ and $\tau = 1.0$ ns.

E. TETRAPYRIDOPORPHYRAZINES

The first tetrapyridoporphyrazines (PyDs) were reported by Linstead et al. in 1937 [96]. In [96], the method of preparation of precursors and PyDs, and the stability and colors of the PyDs are well described, together with some fundamental differences between PyDs and Pcs. Quinolinic cyanamide and etched magnesium turnings were heated with stirring to 270-280°C during 10 min to produce **114**, and from this compound magnesium was eliminated by treatment with sulfuric acid to yield metal-free **115**. Compounds **115** and **116-120** were prepared recently from quinolinic dinitrile (2,3-dicyanopyridine) in the absence and presence of metal salts in N,N-dimethylaminoethanol in 17-74% yield [97, 98] by the "isoindoline" method [95]. Also, the "nitrile" method has been employed [99]. Metal-free **115** was produced by treating the nitrile with lithium in amyl alcohol or in a mixture of trichlorobenzene and benzylalcohol, and some metal complexes were obtained by fusing the nitrile in the presence of metal salts in quinoline or trichlorobenzene at 190-200°C. The yields were fairly high (65-80%) [99]. The starting material, 2,3-dicyanopyridine, is now commercially available, but of course it can be prepared from pyridine-2,3-dicarboxylic acid through its ethyl ester and amide [100, 101]. Some of the metal complexes can also be obtained from pyridine-2,3-dicarboxylic acid, metal salts, and urea, but the purification is more tedious than with the above methods [102-105]. Quarternization of the nitrogens of some of these complexes was tried using methyl iodide or dimethyl sulfate, but the use of the latter alone gave satisfactory results [97, 98, 106-108]. Compounds **121-124** are soluble in water.

Compounds **125-128** are called 3,4-PyDs, since they are synthesized from 3,4-dicyanopyridine, pyridine-3,4-dicarboxylic acid, or the hemiamide of the latter, when needed to distinguish from the above 2,3-PyDs [103-105, 109-111]. The synthetic methods for 3,4-PyDs are essentially similar to those of the 2,3-congeners. Similarly to 2,3-dicyanopyridine [100, 101], 3,4-dicyanopyridine can be prepared from pyridine-3,4-dicarboxylic acid [109]. The copper complex **125** was obtained using both

Tetrapyridoporphyrazine   Structures

114	R=nil	M=Mg
115		M=H₂
116		M=AlOH
117		M=Zn
118		M=Cu
119		M=Ge(OH)₂
120		M=Cd
121	R=(CH₃)⁺(CH₃SO₄)⁻	M=Zn
122		M=H₂, AlOH
123		M=Co
124		M=Cu

125	R₁=H-, R₂=nil	M=Cu
126	R₁=H-,	M=Co, Cu
	R₂=(C₁₈H₃₇)⁺(Br)⁻	
127	R₁=H-, R₂=CH₃-	M=Cu
128	R₁=(CH₃)₃C-, R₂=nil	M=VO, Co

the nitrile [109] and the diacid [103, 110], and the quarternized **126** were obtained by subsequent reaction with octadecyl bromide in refluxing DMF [112, 113]. Compound **127** was obtained similarly as for **124** [109], and *tert*-butylated **128** were synthesized from the hemiamide of 6-*tert*-butylpyridine-3,4-dicarboxylic acid in 18-20% yield [111]. Iwashima and Sawada obtained remarkable results in the preparation of metal-free 3,4-PyD [103]. They reacted pyridine-3,4-dicarboxylic acid and urea at a 1:9 mole ratio at 240°C for 4 h and obtained the desired compound in 59% yield. When this reaction was carried out in the presence of a catalytic amount of ammonium molybdate and at 280°C, the yield was 89%. Similar to the case of 2,3-PyDs, purification is easier if 3,4-dicyanopyridine is used instead of pyridine-3,4-dicarboxylic acid as the starting material. The solubility of 3,4-PyDs in common organic solvents is generally higher than the corresponding 2,3 analogues. In particular, it is noteworthy that many

3,4-PyDs have some solubility in water of both acidic and alkaline pH, although not much at neutral pH [103].

Partial NMR spectra of **121** are shown in [97], where two types of spectra were obtained because of the presence of constitutional isomers. Although not listed, Hanack et al. reported the NMR spectra of several iron complexes of 2,3-PyD [102b]. Many kinds of isocyanide compounds were coordinated as axial ligands, and proton signals from these ligands appeared at upper fields because of the ring-current effect.

IR spectra of **115, 118**, and **125** are shown in [98, 103, 110]. In **125** [110], the bands at 1600 and 1100 cm^{-1} were attributed to C=N. According to [97], **114-120** show a metal-dependent frame vibration at 885-905 cm^{-1}. The IR spectrum of a Langmuir-Blodgett (LB) film of **126** showed a band at 1640 cm^{-1} assignable to a C=N$^+$ stretching vibration [112]. The water-soluble **124** showed bands at 1410, 1200, 960, and 890 cm^{-1}, but when reduced, these bands disappeared, and a new band appeared at 1100 cm^{-1} tentatively assignable to a stretching vibration of the sulfate anion [108].

Electronic absorption spectroscopic data of PyDs are seen in several publications [96-99, 102-106, 108, 110-112, 114]. In general, the longest-wavelength Q bands of the 2,3 isomers lie at 615-670 nm, ~30-50 nm to shorter wavelengths than those of Pcs. Of the 2,3 and 3,4 isomers, the band position of the latter always occurs at longer wavelengths by ~20-50 nm. Accordingly, the Q band position of the 3,4 isomers is close to that of Pcs. The results of molecular orbital calculations also suggest that the Q band of PyDs appears at a shorter wavelength than that of Pcs [111, 115]. Comparing the ε values among Pc, 3,4-, and 2,3-PyDs, they decrease in this order. In contrast, the Soret-band position does not differ significantly among these species. When quarternized, however, the shape of the Soret band changes, as shown on going from **117** to **121** [97]. Formation of an LB film results in a large blue-shift of the Q band [112].

Compound **117** shows a fluorescent peak at 665 nm whose intensity is oxygen independent when excited at 580-642 nm in DMSO [102a].

ESR data have been collected for the cationic species **121, 123, 124** [106-108, 116] and **126** [112, 113]. From these spectroscopic data, it was found that the first reduction of **123** and **124** occurs at the central metals, but that in **121** the ligand is reduced first [108]. ESR spectra of **123** coordinated by isoquinoline, pyridines, or imidazoles are shown in [106], together with observed and simulated ESR parameters. Aggregation properties of **124** were studied by this method [107], and it was found that **124** forms a cofacial dimer in DMF at 77 K and that aggregated **124** in water deaggregates by the addition of DMF at ambient temperature. Salt formation between quarternized 2,3-PyDs and tetrasulfonated Pcs is studied in

[117]. The orientation of **126** molecules in an LB film was determined to be flat on the substrate surface by recording anisotropic ESR spectra and comparing the experimental parameters with theoretical ones [112, 113].

The conductivity and activation energy of **118** are reported in [101]. Over a wide range of temperatures, the conductivity of **118** is higher than that of CuPc, and a value of activation energy of 1.17 eV is reported as compared with 1.3 eV for CuPc.

Cyclic voltammograms of **117** were obtained in DMSO [102]. Irreversible reductions occur at -0.60 and -1.20 V versus SCE, respectively, but no oxidation couple was recorded. When deposited onto a platinum electrode, these reductions were observed to occur at -0.5 and -0.8 V versus SCE in water at pH 6.4, and the layers of **117** behaved like n-type semiconductors having a flat-band potential of -0.38 V versus SCE [118].

Compound **123** was used as a catalyst in the autooxidation of cumol [119], while **124** was used in the hydrocracking of asphaltene [120].

F. TETRAPYRAZINOPORPHYRAZINES

The starting materials of tetrapyrazinoporphyrazine (PyZs) synthesis are 2,3-dicyanopyrazine and its substituted materials [96, 111]. These are easily obtained from diaminomaleonitrile and α-diketones [121]. For example, heating of the former with diacetyl [96] or $tert$-butyl glyoxal [111] in the presence of acetic acid deposited 2,3-dicyano-5,6-dimethyl-pyrazine or 5-$tert$-butyl-2,3-dicyanopyrazine. There is another method leading to 2,3-dicyanopyrazine [101]. Dehydration of 2,3-pyrazine-dicarboxylic acid diamide with phosgene in DMF also produces the nitrile in yields of more than 60%. Synthesis of H_2PyZ **129** was attempted by Linstead et al. [96]. 2,3-Dicyanopyrazine and etched magnesium were heated at 200°C for several hours, and the raw MgPyZ was treated with strong acid, but magnesium was not eliminated completely. Pure **129** was prepared from 5,7-diimino-6H-pyrrolo[3,4-b]pyrazine by the so-called "isoindoline" method in 85% yield [95]. Not only vanadium **130** and cobalt **131** complexes, but also copper and zinc complexes of PyZ were synthesized in 45-80% yield by heating 2,3-dicyanopyrazine and metal or metal salts at 180-190°C for several hours [99]. The iron complex **132** was prepared from the nitrile and pentacarbonyliron in refluxing ethylene glycol [122]. The solubility of **129-132** in the usual organic solvents is not high. The $tert$-butylated compounds, **133** and **134**, soluble in many solvents, were prepared from 5-$tert$-butylpyrazine-2,3-dicarboxylic acid [123], while metal-free **135** was prepared from the corresponding dinitrile [111]. Octaalkylated PyZs **136-138** were synthesized recently [124, 125], and it was concluded that the solubility of **136** in general organic solvents

Tetrapyrazinoporphyrazine Structures

129	$R_1=R_2=H-$	$M=H_2$
130		$M=VO$
131		$M=Co$
132		$M=Fe$
133	$R_1=(CH_3)_3C-, R_2=H-$	$M=VO$
134		$M=Cu$
135		$M=H_2$
136	$R_1=R_2=C_2H_5-, C_3H_7-, C_5H_{11}-$ $C_7H_{15}-, C_9H_{19}-,$ or $C_{11}H_{23}-$	$M=H_2$
137	$R_1=R_2=C_{12}H_{25}-$	$M=H_2$
138	$R_1=R_2=C_{12}H_{25}-$	$M=Cu$

139	$M=H_2$
140	$M=VO$
141	$M=Co$
142	$M=Cu, Zn$

143	$M=VO$
144	$M=Co$
145	$M=Cu$
146	$M=Zn$

is higher than Pcs, and that the tendency to aggregate in chloroform is much less than for Pcs [124].

NMR data of **136** are tabulated in [124]. Signals from two pyrrole protons appear at around -1.2~-1.5 ppm and those from alkyl groups spread between 0.86 to 3.64 ppm because of the ring-current effect of PyZ. Judging from the positions of the pyrrole proton signals, the ring current of PyZ is smaller than that of Pcs, whose signals appear at around -2~-3 ppm [126].

IR data of PyZ are rare, but those of **136** are collected [124]. Eight bands were commonly detected in the following frequency ranges; 3250-3300, 2910-2970, 2850-2870, 1540-1560, 1500-1510, 1455-1470, 1120-1140, and 740-750 cm^{-1}.

UV-visible absorption data are found in several publications [96, 99, 111, 122-125]. The longest-wavelength Q bands appear at around 615-655 nm, shorter by 10-50 nm than in Pcs [23]. Thus, the effective π-conjugation system is smaller than in Pcs. The position of the so-called Soret band does not differ much between Pcs and PyZs.

The half-wave potentials for the first reduction of **137** and **138** were determined to be -0.41 and -0.55 V versus SCE, respectively, in dichloromethane [125]. Since these values are more anodic than those of most Pcs [127], it was concluded that PyZ derivatives are π acceptors.

Both **137** and **138** [125], and perhaps most of **136**, form mesophases (liquid crystals), and the longer the alkyl chains, the lower the phase transition temperature [124]. The electrical conductivity of CuPyZ without any substituent group is higher than both CuPc and 2,3-CuPyD, and the activation energy is 0.81 eV [101]. Maizlish et al. [128] used **131** and **132** as catalysts for the oxidation of *n*-butyl mercaptan by oxygen. These compounds exhibited a considerably higher catalytic activity than CoPc.

Porphyrazine analogues **139**-**142** are called tetra-2,3-quinoxaline porphyrazines, since they are prepared from 2,3-dicyanoquinoxaline [74, 99]. Metal-free **139** was obtained by heating the nitrile [121, 129] in the presence of lithium in a 1:1 mixture of trichlorobenzene and benzylalcohol at 180°C for 1 h [99]. Metallated **140**-**142** were synthesized by treatment of the nitrile with metal or metal salts in di- or trichlorobenzene at 180-190°C for a few hours.

Ring-expanded PyZs **143**-**146** are named tetra-2,3-benzo[*g*]quinoxalinoporphyrazines [130]. The starting material, benzo[*g*]quinoxaline-2,3-dicarbonitrile, was prepared from the corresponding diacid via its diester and diamide in 52% total yield [129]. The synthetic conditions are the same as for **140**-**142**.

No NMR and IR data have been reported for **139**-**146**, but electronic absorption data were collected in [74, 99, 130]. The longest-wavelength Q-band maxima of **139**-**142** lie at 710-770 nm, while those of **143**-**146** are at 785-830 nm. Thus, the Q-band positions of **140**-**142** are just inbetween those of Pcs and 2,3-Ncs, and those of **143**-**146** are shorter than those of 2,3-Acs by ~40-100 nm.

G. ANALOGUES WITH QUINONE UNITS

Structure of Analogues with Quinone Units

147 M = V O
148 M = F e

As analogues classified into this group, only **147** and **148**, known as anthraquinocyanines, have been reported. Freyer treated 2,3-dicyanoanthraquinone, vanadium trichloride, dry urea, and ammonium molybdate at 260°C for 4 h and obtained **147** in 30% yield [74]. It showed a Q_{0-0} band at 758 nm in 1-chloronaphthalene, at a much shorter wavelength than for VOAc (932 nm) in the same solvent. Hiller and Beck utilized **148** as a catalyst for the electroreduction of oxygen [131].

H. UNSYMMETRICAL TETRAAZAPORPHYRINS

Monobenzotetraazaporphyrin **149** was obtained as a byproduct in the preparation of tribenzotetraazaporphyrin **150** [132, 133]. 1,3-Di-imino-isoindoline and succinoimidine were treated in boiling dry ethanol under nitrogen, and after long, tedious column chromatography, **150** was obtained in 5-8% yield. The metal complexes **151** were obtained by metal insertion reaction to **150** in *o*-dichlorobenzene [133]. Bromination of **150** for only 10 min in chlorobenzene afforded **152** as dark blue needles in 79% yield [133]. Compounds **153-155** were obtained in 8-20% yield by ring-expansion reaction of what we call a subphthalocyanine **242** (Fig. 9) [134]. Thus, treatment of **242** with succinoimidine, or the reaction products of 2,3-dicyanonaphthalene or -anthracene with ammonia, in DMSO-chloronaphthalene mixture at 80-90°C gave both the desired **153-155** and the unreacted **242**. However, the separation of the desired compounds from unreacted **242** was simple, since only two colored portions were included in the reaction products, and the color of **153-155** (blue-green) is different from that of **242** (reddish purple). Compared with the mixed condensation method, this ring-expansion reaction has several advantages. Namely, purification is easy, and the yield is relatively high, and most importantly, only monosubstituted-type Pc analogues are obtained. Trials

Unsymmetrical Tetraazaporphyrin Structures

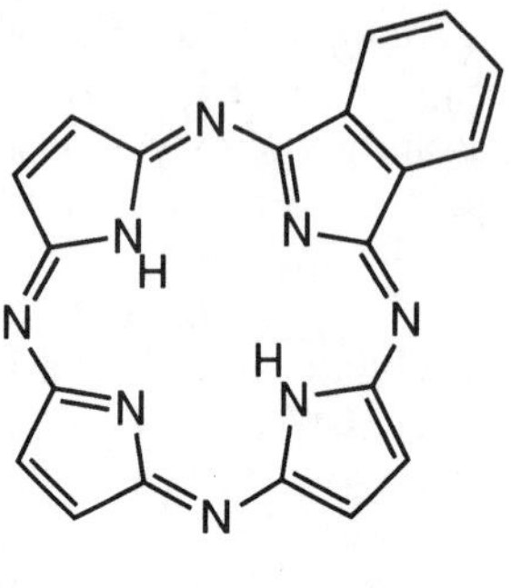

149

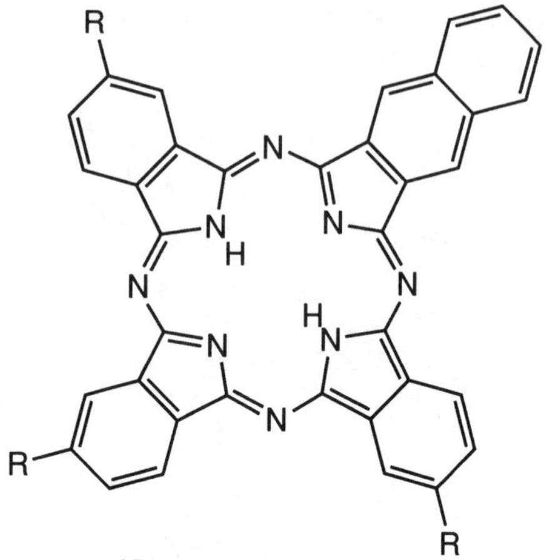

154   R=(CH$_3$)$_3$C-

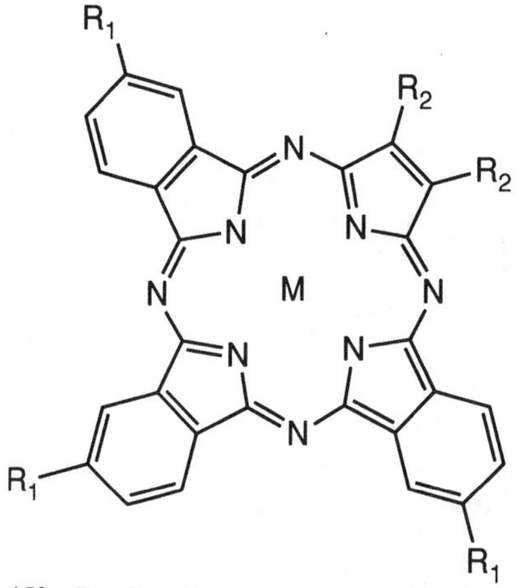

150	R$_1$=R$_2$=H-	M=H$_2$
151		M=Ni, Co
152	R$_1$=H-, R$_2$=Br-	M=H$_2$
153	R$_1$=(CH$_3$)$_3$C-, R$_2$=H-	M=H$_2$

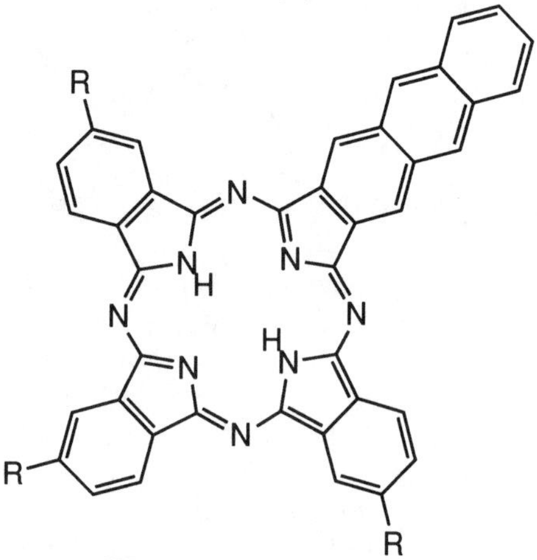

155   R=(CH$_3$)$_3$C-

to optimize the yield were not attempted. However, from many experiments, it was found that the presence of DMSO is very important, although the reaction did not proceed in DMSO alone. Alkoxysubstituted Pc analogues are not readily made by this method. Accordingly, Leznoff et al. synthesized **154** and its neopentoxy analogue by the mixed condensation reaction and subsequent purification using silica-gel columns [134a].

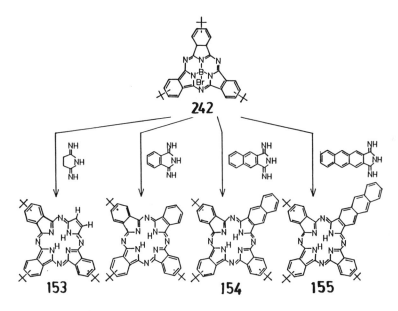

Figure 9 Ring-expansion of a subphthalocyanine (**242**) to various mono-substituted type unsymmetrical Pc derivatives.

No NMR data are available for **149, 150**, and **152**, but data on **153-155** are summarized in [134]. Interestingly, signals from two pyrrole protons appear as three broad peaks due to the difference of attached positions.

Absorption spectra of **149, 150, 152**, and some of their metal complexes are displayed in [133]. Spectra of **149** and its metal complexes indicate that, compared with the spectra of unsubstituted tetraazaporphyrin **6-8** [5], attachment of a fused benzene ring shifts the Q band to longer wavelengths by more than ~50 nm. The spectra of **150-153** are complex in the Q band region because of the low symmetry of the molecules, but commonly show at least two intense bands [133]. Comparison of the spectra among **153-155** indicates that the position of the Q band peaks shifts to longer wavelengths by 20-30 nm per benzene unit [134].

I. TETRABENZOPORPHYRINS

Tetrabenzoporphyrins (TBPs) were first synthesized by Helberger et al. in 1938 [135, 136], and later in 1940-1950 by Linstead et al. [137, 138]. Although several ways to **156** were discovered, the best method at that time

Tetrabenzoporphyrin   Structures

No.	Substituents	M
156	R_1~R_5=H-	M=Zn
157		M=H_2
158		M=Mg
159		M=Ge(OH)$_2$
160		M=Ni
161		M=FeCl
162		M=Fe(Im)$_2$Cl
163		M=Pd
164		M=Pt
165	R_1=R_4=CH$_3$-, R_2=R_3=R_5=H-	M=H_2
166		M=Zn
167		M=Fe
168	R_1~R_4=F-, R_5=H-	M=Zn
169	R_2=(CH$_3$)$_3$C-, other R=H-	M=H_2
170		M=Mg
171		M=Zn
172		M=Cu
173	R_2=Ph-, other R=H-	M=H_2
174		M=Zn
175	R_2=R_3=Ph- other R=H-	M=Zn
176	R_5=CH$_3$-, other R=H-	M=Zn
177		M=H_2
178	R_5=Propyl or Amyl	M=Zn
179	R_5=Ph-, other R=H-	M=Zn
180		M=H_2
181		M=Cd
182	R_5=Ph-, R_2=(CH$_3$)$_3$C-, other R=H-	M=Zn
183	R_5=CH$_3$Ph-, R_1~R_4=H-	M=Zn
184	R_5=CH$_3$OPh-, R_1~R_4=H-	M=Zn
185	R_1~R_4=H-, R_5=1-naphthyl	M=Zn
186	R_1=R_3=H-, R_2=(CH$_3$)$_3$C-, R_5=NO$_2$	M=Zn
187		M=H_2
188	R_1~R_4=H-, R_5=NO$_2$	M=Zn
189	R_1=R_3=H-, R_2=(CH$_3$)$_3$C-, R_5=NH$_2$-	M=Zn
190	R_1=R_3=H-, R_2=(CH$_3$)$_3$C-, R_5=CH$_3$CONH-	M=Zn
191	R_1=R_3=H-, R_2=(CH$_3$)$_3$C-, R_5=C$_{17}$H$_{35}$CONH-	M=Zn
192	R_1~R_4=H-, R_5= [biphenyl]	M=Zn
193		M=H_2
194	R_1~R_4=H-, R_5=Ph & 3H-	M=Zn
195	R_1~R_4=H-, R_5=2Ph & 2H-	M=Zn
196	R_1~R_4=H-, R_5=3Ph & H-	M=Zn

was to heat an intimate mixture of zinc acetate and 3-carboxymethyl-phthalimidine at 340-345°C for several hours under nitrogen. The authors could not completely purify this ZnTBP by chromatography, and it was

finally purified by sublimation. The best recovery was obtained by sublimation at 450°C. In 1978, Vogler and Kunkely prepared **156** by treatment
of 2-acetylbenzoic acid, zinc acetate, and sodium hydroxide at ~400°C for
~1.5 h in the presence of molecular sieves [139]. From the reaction
mixture, **156** was extracted by hot pyridine, and purified using a neutral
alumina column. The starting material, 2-acetylbenzoic acid, is now available commercially but is also easily synthesized in one step in 50% yield
from phthalic anhydride and malonic acid [140]. Luk'yanets et al.
developed another method in 1981 [141]. They reacted phthalimide or its
potassium or sodium salt, malonic acid, and zinc acetate at 350-360°C for 2
h. In these three methods, the yield was 15-26%, but the method of
Luk'yanets et al. gave the highest yield. Remy prepared **156** from isoindoline [142], while isoindoline hydrochloride was utilized by Kopranenkov et
al. [143], both methods giving a very high yield (57-59%). However,
isoindoline is not easily obtainable and is unstable at room temperature
[144]. In addition, the description on isoindoline hydrochloride [145] is not
easily obtained. Demetallation of **156** in concentrated sulfuric acid into
which was bubbled dry hydrochloride gas [138] or in sulfuric acid at 40°C
for 30 minutes [146] produced metal-free **157**.

Koehorst et al. attempted to obtain pure **157,158**, and CdTBP [147].
Pure **157** was obtained by demetallation of CdTBP in a chloroform-tri-
fluoroacetic acid mixture. They obtained the magnesium derivative, **158**,
by metal insertion into **157**, but **158** was also prepared by treatment of 2-
cyanoacetophenone and magnesium in quinoline [135]. Germanium, **159**,
and NiTBP **160** were obtained by metal insertion into **157** in quinoline
[148] or pyridine [149]. FeTBP was obtained by three methods, that is by
using 2-cyanobenzophenone [135] or 2-carboxybenzophenone [146] as
starting materials, or iron insertion into **157** [150, 151]. Treatment of
FeTBP with dilute hydrochloric acid produced **161**, and further treatment
with imidazole yielded imidazole-coordinated **162**. Compounds **163** and
164 were obtained from **157** and palladium chloride or platinum bromide
in refluxing DMF [146].

Octamethyl TBPs (OMTBPs) were generally obtained from 1,3,4,7-
tetramethylisoindole. Not only **165-167**, but also magnesium, manganese,
cobalt, nickel, and copper complexes have been synthesized [152-154]. The
tetramethylisoindole is much more stable than isoindole and could be
prepared in 60% yield in one step from ammonium sulfate and 2,5-hexane-
dione [153], or in 55% yield by acid-catalyzed self-condensation of 2,5-
dimethylpyrrole [155]. The metal complexes were obtained either from
metal acetate and the isoindoline in refluxing trichlorobenzene [152] or by
reaction of metal or metal salts with the indoline in the absence of solvents
[153]. The yields were between 25 and 83%, depending on the kind of
metal and procedure. Metal-free **165** was obtained from the manganese

complex by the action of trifluoroacetic acid [153]. Reaction of **165** with metal salts in quinoline or DMF also produced MtOMTBPs [153]. Mt-OMTBPs have a dark-blue color and are more soluble than MtTBPs [153]. Hexadecafluorinated **168** was prepared [142] from 4,5,6,7-tetrafluoroiso-indole [156], zinc acetate, and formaldehyde in 57% yield. *Tert*-butylated TBP **169** and its magnesium **170**, zinc **171**, copper **172**, nickel, manganese, vanadium, iron, and cobalt complexes were prepared from 3-carboxymethyl-5-*tert*-butylphthalimidine, and purified by chromatography [157]. The phthalimidine was prepared from 6-*tert*-butyl-2-naphthol in 22% yield via four steps. However, later, a method using *tert*-butylphthal-imide was introduced [141]. This phthalimide is easily obtained in one step from commercially available *tert*-butylphthalic anhydride and urea in 60-80% yield. Since the yields of the TBPs are not very different between the two methods, the latter method is preferable. Compounds **173-175** were prepared using this latter method [158]. Thus, the substituted phthalimides or their potassium derivatives and malonic acid or sodium acetate and zinc acetate were heated under a nitrogen atmosphere at 360-400°C for 1-2 h, to afford the zinc complexes **174** and **175** in yields of up to 22%. Introduction of hydrogen chloride gas into a chloroform solution of **174** produced metal-free **173**. Although not listed here, [158] also describes the synthesis of tetra-methylated or -methoxylated ZnTBPs.

Meso-tetraalkylated TBPs, **176-178**, were reported in 1983 [159]. Tetramerization of 3-alkyl-methylenephthalimidines with zinc acetate in the presence of tribenzylamine at 350-360°C for 30 min in an inert atmosphere produced the zinc complexes **176** and **178** with yields of ~14%. In the absence of tribenzylamine, the yield was zero or very small. Demetallation of **176** in chloroform by hydrogen chloride gas led to metal-free **177**. These compounds were more soluble than TBPs having no substituent groups, and accordingly, could be purified using alumina columns. Zinc complexes of *meso*-tetraphenyl TBP **179** were prepared in several ways. Luk'yanets's group heated a mixture of potassium phthalimide, sodium phenylacetate, and zinc acetate at 350-360°C for 30 min, and obtained **179** in 17% yield [160]. When a mixture of isoindoline hydrochloride, tribenzylamine, and zinc acetate was treated similarly, **179** was obtained in 10% yield [143]. Remy utilized isoindoline, benzaldehyde, and zinc acetate, and attained a 50% yield [142]. From the standpoint of the yield, Remy's method is superior to the other two methods. However, since isoindoline is unstable at room temperature and rather difficult to obtain, the first method may be recommended. Metal-free **180** can be obtained by dissolving **179** in acetic acid and allowing it to stand for 30 min at 20°C [160]. The cadmium complex, **181**, was obtained as for **179**, using cadmium acetate instead of zinc acetate [160]. ZnTBP containing both *tert*-butyl and phenyl groups, **182**, is also known [161]. Although the synthetic procedure is not reported, this compound would be obtained as for **179**, using *tert*-butylated phthalimide or its potassium salts. Tetra-(*p*-tolyl)-, -(*p*-methoxyphenyl)-, and -(1-naphthyl)ZnTBPs, **183-185**, were prepared as for **179** in yields of up to 13% [160].

Nitration of ZnTBP **156** and its *tert*-butylated compound **171** was carried out using 5% nitric acid in an acetic acid-trifluoroacetic acid mixture or in acetic acid at room temperature [162]. From a mixture of nitrated compounds, **186** and **188** were obtained in 60 and 17% yields, respectively, after repetitive chromatography. Demetallation of **186** using acetic acid and hydrogen chloride gas (25°C, 30 min) gave **187** in 88% yield. Reduction of the nitro groups of **186** with tin in acetic acid produced crude **189**, and subsequent treatment of **189** with acetyl chloride gave **190** in 78% yield. Stearylamino-group-attached **191** was obtained in 40% yield when **189** and stearyl chloride were treated in refluxing benzene for 1 h [162]. *Meso*-tetrabiphenyl ZnTBP, **192**, was synthesized [163] from sodium biphenylacetate, zinc acetate, and the potassium salt of phthalimide by the same procedure as for **179** [160]. Demetallation of **192** in an acetic acid-hydrochloric acid (4:1 v/v) mixture (reflux of several hours) furnished **193**. *Meso*-phenyl ZnTBPs containing various numbers of phenyl groups, **194-196**, were reported recently [164a]. Ichimura et al. reinvestigated the reaction of 3-benzylidenephthalimidine with zinc acetate, which was introduced previously for the preparation of **179** [160]. From the reaction mixture, they obtained **179**, **195**, and **196** in 0.4, 5.4, and 4.9 %, respectively, together with some amounts of **194** [164a]. Their recent publications [164b] indicated that the use of zinc benzoate as a chelating agent produced only tetraphenylated **179**.

X-ray analyses of the structures of NiTBP **160** [149] and NiOMTBP [165] were carried out recently. A perspective view of NiTBP suggests that this molecule is as flat as Pcs [166], but the introduction of eight methyl groups at the 3 and 6 positions deforms the structure to some extent [165].

IR data on TBPs cannot be found easily, but those of ZnTBP **156** are known [167]. Signals were observed at 3000 (C-H, weak), 1500-1700 (aromatic nucleus), and 695, 740, and 755 cm^{-1} (strong, aromatic deformations).

^{1}H NMR data are reported for several TBP complexes. *Meso*-protons of Zn-, H$_2$-, and MgTBPs (**156-158**) and CdTBP appear at around 11.3 ppm, while 3,6 and 4,5 protons are detected as double doublets at around 9.9 and 8.3 ppm, respectively [147, 168]. When all *meso* protons are replaced by nitro groups, 3,6 and 4,5 protons appear at 9.3 and 8.1 ppm, respectively. In tetraphenylated ZnTBP **179**, 3,6 and 4,5 protons are shifted to 7.5 and 7.4 ppm, respectively, and protons from phenyl groups appear between 7.9 and 8.6 ppm [164]. The NH proton signal of **180** appears at -1.12 ppm [164b]. NMR spectra of FeTBP coordinated by isocyanides have been reported [151]. By coordination, proton signals of isocyanides shifted to upfield by 1.1-2.6 ppm because of the ring current of TBP.

Electronic absorption spectra of TBPs are reported in a number of publications [5, 7, 137-139, 141, 143, 146-150, 152, 157-160, 162-164, 166-175]. There are several characteristic features in the spectra of TBP. (1) The position and intensity of the Q bands are inbetween those of Pcs and general porphyrins. (2) In contrast to Pcs and general porphyrins, the Soret band of metal-free TBP shows split x and y components [5, 7]. (3) The shape of the Q band of MtTBPs is close to that of MtPcs. The small absorption peak often seen at the longer-wavelength side of the Soret band (450-460 nm) in old literatures is due to an impurity that is difficult to remove [147]. In [147], this band is tentatively ascribed to a partly hydrogenated TBP. Spectra of some TBPs are also reported in n-octane at 4.2-6 K [176, 177]. At such low temperatures, a lifting of the degeneracy is induced by the action of the crystal field on the Jahn-Teller unstable S_1 state, so that the Q bands of several TBPs are split by ~20-40 cm^{-1}. The spectra of oxidized TBPs are reported for MgTBP **158** [172b], ZnTBP **156** [171], FeTBPCl **161** [150], and FeTBP(CO)(Py) [178]. In all cases, the first oxidation occurs at the TBP skeleton, and both the Soret and Q bands show hypsochromic shifts and hypochromic effects, although three peaks are additionally observed at the longer-wavelength side of the Q band of MgTBP **158** [172b]. The spectra of reduced TBP species are rare, but are known for ZnTBP **156** [30a] and FeTBPCl **161** [150]. The Q-band peaks of mono- and dinegative ZnTBP appear at shorter or longer wavelengths, respectively, to the Q band of the neutral species in accord with those of the skeleton-reduced Pcs [30a]. Although the iron in FeTBPCl **161** is trivalent, reduction of this species displayed spectra ascribable to Fe(II) high-spin and subsequently Fe(I)TBP species [150].

Concerning MCD spectra of TBPs, those of ZnOMTBP **166** [175] and FeTBP derivatives, **161** and **162** [150], were recorded. The MCD spectra of **166** changed markedly, depending on both temperature and concentration. In addition to the spectrum corresponding to monomeric species, three types of spectra were recorded, and these were attributed to three different aggregate structures. The spectrum of **161** [150] was that of Fe(III) high-spin porphyrins [179, 180]. Thus, in addition to Faraday A terms associated with the main Soret and Q bands, another Faraday A term was detected in one of the near-IR charge-transfer (CT) bands. The bis-imidazole complex **162** displayed a Q-band MCD whose intensity was proportional to the inverse of the absolute temperature (Faraday C terms), indicating that the compound was in the Fe(III) low-spin state.

The luminescence properties of H$_2$-, Zn-, Mg-, Pt- (**156-158** and **164**), and CdTBPs, and some lanthanum complexes of TBP have been examined intensively, especially by Russian workers [169, 181-189] and by Gouterman et al. [173, 190]. In contrast to general porphyrins and Pcs, several (Zn, Cd, and Lu) TBP derivatives show S_2 emission [187, 190], and even the presence of S_3 emission is suggested for H$_2$TBP **157** [183, 188, 189]. The quantum yields of the S_1 fluorescence of ZnTBP **156** are 0.35 in octane with 0.5% pyridine by volume [190], 0.22 in toluene [184], 0.10 [187] or 0.11 [191] in DMF, and those of S_2 fluorescence are 0.0016 in octane with 0.5% pyridine by volume [190] and 0.0013 in DMF [187]. The quantum yield of S_1 emission of MtTBP increases upon changing the central metal in the order of Zn<H$_2$<Mg, similarly to the Pc series [187]. Comparing TBP and Pc with the same central metal, the quantum yields in Pc appear larger [187]. At 4.2-20 K, the emission spectra of Zn-, Mg-, and CdTBP show splitting [177, 190, 192]. Pt(II)TBP shows strong phosphorescence at 745 nm, with a quantum yield in degassed pyridine at room temperature of 0.8 [173].

Theoretical molecular orbital calculations on TBP molecules have been carried out by several researchers [27, 35, 37, 193-195]. In contrast to general porphyrins, where the a_{1u} and a_{2u} orbitals are accidentally degenerate, the energies of the a_{1u} and a_{2u} orbitals are different in TBP, with the former being higher. However, their energy difference is smaller than in tetraazaporphyrin and Pc systems. According to Lee et al. [37], the Q band primarily involves a transfer of charge from the pyrrole carbon skeleton to the *meso*-carbon and outer benzene rings, and the $a_{1u}, a_{2u} \rightarrow e_g$ excitations comprise only 60% of the Soret (B) band. In [193], counter maps of the two-dimensional wave functions at various energy levels are shown, which is very conv enient for understanding the shape of the orbitals visually. Solov'ev et al. [194] discussed the influence of azo-substitution of *meso*-carbons of MtTBP on the absorption and luminescent spectra within the framework of the LCAO method. One of the most important results was that *cis*-type substitution by two nitrogens does not produce a split Q band, while *trans*-type substitution produces, at least theoretically, a split Q band. This indicates further that *trans* D_{2h} porphyrins may show a split Q band, but that *cis*-type C_{2v} porphyrins do not. This was recently confirmed using Pc derivatives experimentally and on the basis of a molecular-orbital calculation using Pariser-Parr-Pople method [69]. A model was developed to explain the observation that several prominent bands in the absorption spectrum of MgTBP **158** have no counterpart in the emission spectrum [195]. The model led to simultaneous destructive interference between displacement (Franck-Condon or Jahn-Teller) and vibronic-coupling (Herzberg-Teller) contributions to the intensity of the missing bands.

ESR spectra of oxidized [196] and reduced [197] ZnTBP **156** clearly indicated that the HOMO of this compound has a_{1u} symmetry. That is, since the a_{1u} orbital has nodes at the pyrrole nitrogens, the 9-line spectrum due to four $^{14}N(I{=}1)$ was absent. The ESR of copper, cobalt, and vanadyl TBPs and OMTBPs was examined minutely by Lin et al. [198, 199]. ESR parameters of not only TBPs, but also several porphyrins and Pcs, are tabulated and discussed in [198], and in [199] it is shown that the relative d-orbital energies of cobalt porphyrin systems can be deduced from their ESR data by assuming values of the Racah parameters and the spin-orbit coupling parameter, as in Fe(III) low-spin porphyrin systems [200]. Zeroth-order (ligand field plus electrostatic) energies of the multiplet states for the various cobalt porphyrin systems are shown in a figure. The energy order of the multiplet states changes dramatically according to the type of porphyrin skeleton. That the first oxidation of ZnTBP **156** and FeTBP(CO)(Py) occurs at the TBP ring was confirmed by the ESR technique [171, 178]. A single line having a width of 5.0-5.5 G and a g value of 2.003 was observed in these systems. ESR was applied to FeTBPCl **161** in order to identify the spin and oxidation states of iron in this complex [150]. Iron in this compound was in a high-spin trivalent state. When imidazoles were coordinated to **161** to form mono and bisadducts, iron in these compounds converted to a low-spin trivalent state. Ligand fields in these complexes were analyzed from the g values according to [200]. Both the tetragonal and rhombic splitting values were smaller than those of general Fe(III) low-spin porphyrin systems, suggesting that the TBP skeleton is more planar in general than porphyrin skeletons. Koehorst et al. obtained zero-field splitting parameters (D and E) of MgTBP **158** and its mono- and bis-pyridine adducts, and H$_2$TBP **157** from their triplet ESR spectra [147]. They showed that at high temperature, $E{=}0$, whereas at low temperature, $E{\neq}0$. This indicates that the two mutually perpendicular in-plane axes (x and y) of MgTBP **158** become nonequivalent at low temperatures.

Mössbauer spectroscopy was applied to FeOMTBP **167** in order to determine the electric ground state of iron [201]. At 4.2 K, the electric field gradient at Fe is positive and axially symmetric, consistent with a spin quintet $^5B_{2g}$ ground term. But recently, Smirnov et al. proposed an ^5E_g term [202]. The bistetrahydrofuran adduct of this compound was also high-spin, whereas the bispyridine adduct was diamagnetic with $^1A_{1g}$ ground state [201]. Sams et al. measured the Mössbauer spectra of the *trans*-dicarbonyl derivative of **167** [203]. The principal component of the electric field gradient tensor at Fe was negative in this complex, reflecting the weaker equatorial ligand field of OMTBP.

Coordination of bases to ZnOMTBP **166** was examined and compared with that of Zn tetraphenylporphyrin (ZnTPP) [204]. The equiliblium constants for the formation of monoadducts were 10-100 times greater than those in ZnTPP systems. It was claimed that this was indicative of an extremely electron-deficient metal center, and therefore that the OMTBP ligand is an extremely weak σ donor. The equilibria and kinetics between bisadducts of FeOMTBP **167** and carbon monoxide were studied in toluene [174]. Equilibrium constants for the formation of mono-carbonyl adducts were intermediate between those found for the corresponding general porphyrin systems on the one hand and Pc on the other.

Some redox potentials of ZnTBP **156** [30b, 171], H2TBP **157**, and FeTBPCl **161** [150] were collected. ZnTBP is oxidized reversibly at 0.36 and 0.97 V versus SCE and reduced at -1.48 [171], -1.84, -2.49, and -2.70 V in DMF [30b]. The potential difference of 1.84 V found between the first ring oxidation and reduction of TBP is much lower than that observed for general porphyrin systems, ~2.25 V [205]. FeTBPCl **161** shows Fe(III/II) and Fe(II/I) redox couples before oxidation and reduction of the TBP skeleton in DMF, DMSO, and pyridine. The potential difference between the first ring oxidation and reduction in this compound is 2.3-2.4 V, suggesting significant interaction between iron and the TBP ligand.

The solution conductivity of FeTBPCl **161** was measured at various concentrations [150]. It was found that chlorine coordinated at the axial position dissociates much easier than in the FeTPPCl system. This fact indicates that FeTBPCl can be placed midway between FeTPPCl and FePc, because if chlorine dissociates and moves away from FeTBPCl, the oxidation state of iron becomes divalent, which is the oxidation state of iron in FePc. The powder dc dark conductivity of CoTBP coordinated by one iso-thiocyanate was 6×10^{-5} S/cm [206].

Several MtTBPs have been used as catalysts for the electroreduction of oxygen [207]. Heat treatment of CoTBP in an inert gas at 800-900°C improved its stability [208].

Zn- and MgTBPs have been the subject of photochemical hole burning [209, 210] and photovoltaic cell [211] studies.

Although not touched upon here, MtTBPs and some of their oxidized derivatives form one-dimensional molecular stacks, as reported for Pcs [212]. Since these compounds are outside the scope of this chapter, the reader is referred to the following references as representative in this field [148, 149, 151, 154, 165, 213-217].

J. TETRANAPHTHALOPORPHYRINS

Compounds **197** and **198** are called tetra-1,2-naphthaloporphyrins (1,2-TNPs). Zn-1,2-TNP **197** was prepared in 25% yield by heating a mixture of potassium 1,2-naphthalenedicarboximide, sodium acetate, and

Tetranaphthaloporphyrin   Structures

| 197 | M=Zn |
| 198 | M=H$_2$ |

199	R$_1$=R$_2$=H-	M=Zn
200		M=H$_2$
201		M=Fe(Py)$_2$
202		M=Co(II)
203		M=Co(III)Cl
204		M=Co(III)CN
205	R$_1$=Ph-, R$_2$=H-	M=Zn
206	R$_1$=H-, R$_2$=Ph-	M=Zn
207		M=H$_2$

zinc acetate at 390-400°C, and treatment of **197** with hydrogen chloride gas in DMF furnished **198** in 25% yield [218]. Electronic absorption spectra of these compounds do not differ significantly from those of ZnTBP **156** and H$_2$TBP **157** in the shape, intensity, and positions. As mentioned in the section on naphthalocyanines, the absorption spectra (especially positions) change markedly when π-conjugated systems are expanded radially, as seen from the center.

Zn-2,3-TNP **199** was originally prepared in 12% yield by treatment of 3-(carboxymethyl)-5,6-benzophthaloimidine and zinc acetate at 340-360°C [219], and later two other methods were tried [218]. The phthal-imidine was prepared in ~60% yield from 1-nitroso-2-anthrol, via a two-step reaction. Demetallation reaction of **199** in a chloroform-concentrated hydrochloric acid mixture [218] or in trifluoromethane sulfonic acid in an inert atmosphere at room temperature for 15 min [220] furnished **200** in ~80% yield. Reaction of 1-(2,3-dihydro-1-hydroxy-3-oxo-(1*H*)-benz[*f*]iso-indole) acetic acid dihydrate with freshly prepared iron(II) acetate at 350°C under a nitrogen atmosphere, and subsequent extraction of the product with pyridine gave **201** in 3% yield [220]. Heat treatment (170°C, 3 h in vacuum) of **201** removed two pyridine molecules. A cobalt insertion reaction on **200** in pyridine and subsequent acid treatment of the product yielded **202** in 33% yield. Oxidation of **202** by thionyl chloride in nitrobenzene or by potassium cyanide in ethanol produced **203** in 78% or **204** in 85%, respectively [220]. The tetraphenylated compound **205** was obtained by the reaction of 1-phenyl-2,3-naphthalenedicarboximide, sodium acetate, and zinc acetate at 350-360°C for 1-2 h. Similar treatment of 2,3-naphthalenedicarboximide or its potassium salt, phenyl acetic acid, and zinc acetate gave **206** in 25% yield. Demetallation of **206** by hydrogen chloride gas in chloroform produced **207** in 73% yield [218].

NMR data have been collected for **201** [220]. The three different protons on the naphthalene ring, constituting eight protons each, appeared at 7.84, 8.68, and 10.16 ppm in pyridine. The more inner the position, the lower the field, because of the ring-current effect of TNP. The four *meso*-proton signals appeared at 11.42 ppm.

IR data on **201-204** are described in [220].

Electronic absorption spectra of **206** and **207** are shown in [218]. Compared with the spectra of the related TBP compounds [160], the Q bands are intensified and shifted to longer wavelengths by ~50-60 nm, while the Soret bands are shifted by 35-45 nm. The Q and Soret bands of **201** occur at 684 and 421 nm in pyridine, and those of **202** at 689 and 425 nm, respectively [220]. Figure 10 shows the absorption and MCD spectra of the demetallated derivative of **205**. Including this species, the splitting of the Q_x and Q_y bands in metal-free TNPs (ca. 200-400 cm^{-1}) is much smaller than that in metal-free TBPs (ca. 900-1300 cm^{-1}) [5, 7] or in general porphyrins (ca. 2000-3000 cm^{-1}), as seen in ring-expanded metal-free tetraazaporphyrin compounds (Figs. 7 and 8). Accordingly, the shape of the MCD curve associated with the main Q absorption band looks like Faraday A term. Interestingly, three or four absorption peaks are generally observed in the Soret region of the non-metallated TNPs [4, 218], although metal-free TBPs and usual porphyrins show two [5, 7] or one Soret absorption peak(s), respectively. The ϕ_F of S_1 emission in toluene is 0.14.

Polycrystalline **202** showed two ESR signals, at g=2.42 and 1.99 [220], which are typical for cobalt porphyrins and Pcs in the low-spin, square-planar configuration [221]. Similarly, polycrystalline **203** showed signals at g=2.00, 2.47, and 2.0028, and **204** displayed only one signal at g=2.0031.

Powder dc dark conductivities of **199-204** and some of their oxidized forms are tabulated together with those of MtPcs in [220]. The value for **200** is four orders of magnitude larger than that of H$_2$TBP, indicating that the enlargement of the π-conjugated system dramatically increases the conductivity. In the absence of axial ligands, cobalt complexes have higher

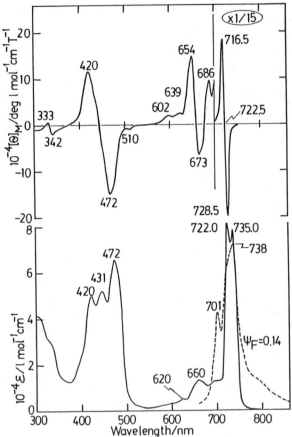

Figure 10 MCD (top) and absorption and fluorescence (bottom) spectra of the demetallated derivative of **205** in toluene. In fluoresence spectrum, the excitation was at 650 nm.

conductivity than iron and zinc complexes by one order of magnitude. Cobalt(III) complexes have much higher values compared with cobalt(II) complexes, especially the value for **203**, which amounts to 9 x 10^{-2} S/cm.

K. TETRAANTHRAPORPHYRIN

Zinc tetraanthraporphyrin (ZnTAntP) **208** is the only TAntP we are aware of at present. It was synthesised from 2,3-anthracenedicarboximide, sodium *p*-biphenyl acetate, and zinc acetate by heating at 350-360°C [4]. The *Q* and Soret bands occur at 772 and 465 nm in dioxane, respectively.

Tetraanthraporphyrin   Structure

Compared with the spectra of *meso*-tetraphenylated ZnTNP **206**, these peaks are shifted to longer wavelength by ca. 75 and 30 nm, respectively. According to the calculation of Solov'ev et al. [222], the Q band of MtTAntPs have been anticipated at around 800 nm.

208   R=

L. MIXED ARENOPORPHYRINS

Mixed unsymmetrical arenoporphyrins **209-216** have all been reported in the literature [223]. Zinc derivatives are obtained by mixed condensation of substituted or unsubstituted phthalimides and 2,3-naphthalenedicarboxiimides or their potassium derivatives in the presence of sodium acetate and zinc acetate at 340-360°C, and subsequent isolation of the products by column chromatography. The yield of the products depends markedly on the ratio of the starting materials. Demetallation of the zinc complexes was carried out by passing hydrogen chloride gas through a benzene solution for 1 h at 20°C. Some zinc complexes, particularly **213**, were unstable, so that reducing reagents such as triethylamine and ethanol had to be mixed as an eluent in chromatography. Compounds **217** and **218** were obtained by mixed condensation of 2,3-naphthalenedicarboximide and tetraphenylphthalimide [224]. First, the zinc complex was synthesized; however, since it was unstable, **217** was isolated after demetallation treatment. Insertion of copper to **217** was conducted in DMF to produce D_{2h}-type **218**.

NMR spectroscopic data are given for **209**, **210**, and **217**. In **209**, *meso*-protons appeared at 12.41, 12.66, and 12.94 ppm, while *tert*-butyl protons gave two multiplets at 1.81-1.84 and 1.94-1.98 ppm. In **210**, signals from pyrrole protons were found at -2.47, -2.49, -2.52, and -2.57

Mixed Arenoporphyrin Structures

209	R$_1$=(CH$_3$)$_3$C-, R$_2$=H-	M=Zn
210		M=H$_2$
211	R$_1$=H-, R$_2$=Ph	M=Zn
212		M=H$_2$

213	R$_1$=H-, R$_2$=(CH$_3$)$_3$C-	M=Zn
214		M=H$_2$
215	R$_1$=Ph-, R$_2$=H-	M=Zn
216		M=H$_2$

217	R=Ph-	M=H$_2$
218		M=Cu

ppm. These four signals at high field have been attributed to the asymmetry of the molecule rather than the existence of isomers with different orientations of the *tert*-butyl group. The *meso*-proton signal of **217** appeared at 10.21, 9.78, 9.64, and 9.27 ppm in chloroform-DMSO 1:1 v/v mixture, suggesting that the symmetry of **217** is probably S_4. The pyrrole proton signal was observed at -1.77 ppm.

Electronic absorption spectra of some of **209-216** are shown in [223]. With the enlargement of the π-conjugate system, both the Q and Soret bands shift to longer wavelengths, and the relative intensity of the Q band to the Soret band increases. Because of the low symmetry of the molecules, the longest-wavelength Q band splits into at least two peaks even in zinc complexes. D_{2h}-type **218** showed the main Q and Soret bands at 677 and 440 nm in methylene chloride [225].

M. TETRAARENOAZAPORPHYRINS

Tetrabenzomonoazaporphyrins **219-224** were synthesized in England [137], Germany [226], and Russia [227, 228]. Barrett et al. [137] obtained **219** by two methods. First, an intimate mixture of 1-imino-3-dicarboxymethylenephthalimidine and zinc dust was heated at 330-340°C for 20 min, and the residue was washed by acid and crystallized from quinoline to give **219** in 27% yield. In the second method, methylmagnesium iodide and powdered phthalonitrile were heated at 200°C for 5 min, and the resultant residue was treated with acid to give **219** in 17% yield. Metal insertion into **219** in boiling quinoline furnished **220-222**. The copper complex alone was also prepared from *ortho*-halogenoacetophenone and cuprous cyanide by heating them in quinoline at 210-220°C [226]. Cadmium and palladium complexes **223** were synthesized from a mixture of carboxymethylenephthalimide, *ortho*-cyanobenzamide, and cadmium or palladium acetate by heating for 3 h at 340-350°C [227]. *Tert*-butylated **224** and tetrabenzo-*trans*-diazaporphyrin **225** were separated from condensation products of potassium *tert*-butylphthalimide, malonic acid, phthalimide, or 3-iminophthalimide, and a compound in which two isoindoline molecules are connected by a methine bridge [228].

Tetrabenzotriazaporphyrins **226-228** were prepared in the late 1930s [229, 230], while **229-234** were synthesized recently [231]. Methylmagnesium iodide and phthalonitrile were reacted first without heating, then at 200°C, and the residue was treated with acid to furnish **226** in 40% yield. Metal insertion into **226** in quinoline produced **227** and **228**. The copper complex alone was obtained also by fusion of *ortho*-cyanoacetophenone and phthalonitrile in the presence of cuprous cyanide [230]. Compounds **229-234** were obtained in an essentially similar manner as for **226** using Grignard reagents and substituted phthalonitriles [231]. If the final acid treatment is eliminated, magnesium compounds are obtained. A tetra-

Tetraarenoazaporphyrin *Structures*

219	R=H-	M=H₂

219	R=H-	M=H$_2$
220		M=Zn
221		M=Mg
222		M=Cu, Fe
223		M=Pd, Cd
224	R=(CH$_3$)$_3$C-	M=H$_2$

226	R$_1$=R$_2$=H-	M=H$_2$
227		M=Zn, Mg
228		M=Cu, Fe
229	R$_1$=(CH$_3$)$_3$CCH$_2$O-, R$_2$=C$_3$H$_7$-	M=Mg
230	R$_1$=(CH$_3$)$_3$CCH$_2$O-, R$_2$=C$_{15}$H$_{31}$-	M=H$_2$
231	R$_1$=(CH$_3$)$_3$C-, R$_2$=C$_{15}$H$_{31}$-	M=H$_2$
232	R$_1$=H-, R$_2$=C$_{15}$H$_{31}$-	M=Mg
233	R$_1$=(CH$_3$)$_3$C-, R$_2$=Ph-	M=H$_2$
234	R$_1$=H-, R$_2$=Ph-	M=Mg

225

235 R$_1$=(CH$_3$)$_3$C-, R$_2$=C$_{15}$H$_{31}$-

naphthalotriazaporphyrin **235** was obtained similarly using *tert*-butyl-naphthalonitrile and hexadecylmagnesium chloride [231]. The solubility of azaporphyrins decreases with increasing number of *meso*-nitrogens if comparison is made among azaporphyrins having no substituent groups [137].

X-ray structural determination was performed for **219** [232, 233] and **226** [234]. As expected, their structure is similar to that of metal-free Pc [166].

NMR spectra were reported for **230, 231, 233,** and **234** [231]. Two pyrrole protons appear as multiplets at -4.30~-1.6 (**230** and **231**) and at -0.45~-0.27 (**233**) ppm. Thus, the ring-current effect is greatly reduced when the phenyl group is attached to the *meso*-carbon. Borovkov et al. [235] claimed from NMR results that the position of pyrrole protons in a tetrabenzotriazaporphyrin changes with temperature and the kind of solvent.

Absorption spectroscopic data of **219-234** are found in [1, 137, 194, 229, 231]. Generally, the Q band intensifies and shifts to longer wavelengths with increasing number of *meso*-nitrogens [137]. Accordingly, the Q-band positions of tetrabenzotriazaporphyrins lie at shorter wavelengths than those of Pcs, while the intensity in the Soret region decreases with the introduction of *meso*-nitrogen. Because of the low symmetry of the molecules, many small peaks are observed in both the Soret and Q-band regions.

Luminescence properties of **219-221** and **223-227** are reported in [194, 236]. All compounds show the so-called S_1 emission, and **220** alone also exhibits S_2 emission [236].

Solid-state properties of oxidized one-dimentional metallotetrabenzo-triazaporphyrins have also been studied extensively. For those who are interested in this field, see, for example [237-239].

N. SUPERPHTHALOCYANINES

A superphthalocyanine (SPc) **236** was synthesized as early as 1964 [240], and its structure was confirmed by x-ray crystallography in 1975 [241]. Phthalonitrile and absolutely anhydrous uranyl chloride were heated in DMF at 170°C for 30-80 min. From the reaction mixture, SPc was isolated in 24% yield. The use of other uranyl salts and the presence of moisture decreased the yield significantly [242]. In a similar manner, alkylated SPcs **237-239** were prepared using substituted phthalonitriles, but in very low yields [243].

X-ray structural analysis of **236** revealed a structure severely and irregularly distorted from planarity, and this was attributed to steric strain within the macrocycle [241]. The distance of U-N contact observed in **236** is estimated to be 2.5-2.6 Å, as opposed to 1.85-2.0 Å for Mt-N in Pcs [244].

Superphthalocyanine Structures

236 $R_1=R_2=H-$
237 $R_1=CH_3-, R_2=H-$
238 $R_1=R_2=CH_3-$
239 $R_1=R_2=C_4H_9-$

IR and far-IR data of **236** are tabulated in [240], together with those of several Pcs. Several bands that were not seen in Pcs were observed, and a band at 278 cm^{-1} was assigned to an O-U-O bending frequency.

NMR spectra of **236** and **237** are shown in [242], and data on **236-239** were summarized in [243]. Because of the distorted structure and ring-current effect, aromatic protons in **236** and **237** appeared as a multiplet in low fields.

The Q- and Soret-band peaks of SPcs appear at 910-940 and 415-425 nm in 1-chloronaphthalene, respec-tively, as a consequence of the large π systems.

SPcs undergo a ring-contraction reaction to Pcs in the presence of many transition metal salts [242].

O. SUBPHTHALOCYANINES

The group of compounds **240-242** are conventionally called sub-phthalocyanines (SubPcs), in contrast to SPcs, which contain five isoindole rings. Meller and Ossko heated a mixture of phthalonitrile and phenyl-difluoroborane or phenyldichloroborane in 1-chloronaphthalene at 260°C for 10 min, and obtained **240** and **241** in ~20 and 40% yield, respectively [245]. Similarly, **242** was obtained in ~50% yield from *tert*-butylphthalo-nitrile and phenyldibromoborane or borane tribromide [246].

The structure of **241** was determined in 1974 [247]. It has a cone-shaped bent structure because of the constraint arising from three iso-indoline units.

500 MHz ^{1}H NMR data of **242** in chloroform are reported in [246]. In comparison with those of tetra-*tert*-butylated H$_2$Pc in the same solvent, the corresponding proton signals appear at higher fields.

In IR spectra, the vibrations associated with B-F, B-Cl, and B-Br bonds occurred at 1061, 950, and 890 cm^{-1}, respectively.

Subphthalocyanine Structures

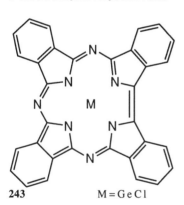

240   R=H-   M=BF
241   M = B Cl
242   R=(CH₃)₃C-   M=BBr

Absorption, MCD, and fluorescence spectra of **242** are shown in Fig. 11, together with those of its planar binuclear dimer **246**. The absorption coefficients at the Q and Soret bands are smaller than those of Pcs [23, 246], and the position of the Q band is much shorter than that of Pcs. Accordingly, SubPcs have a reddish purple color, different from the blue-green color of Pcs.

When heated in the presence of isoindoles, SubPcs produce monosubstituted-type Pcs and Pc analogues, such as **153**-**155** (Fig. 9) [134].

P. TETRABENZOTRIAZACORROLE

Tetrabenzotriazacorrole Structure

243   M = GeCl

A tetrabenzotriazacorrole (TBC) **243** was prepared by reacting GePc(OH)₂ with hydrogen selenide in quinoline in 82% yield as a dark-green powder [248].

IR bands observed at 3250 and 718 cm⁻¹ were attributed to GeO-H and Ge-OH, respectively. In NMR spectra, eight internal and external protons of the TBC ring appeared as multiplets at 10.2 and 9.1 ppm, respectively. Although the shape of the Q band is similar to that of Pcs, it lies at a slightly shorter wavelength (peak at ~660 nm) than in Pcs. A very sharp Soret band exists at 440-450 nm, and its ε is almost twice as intense as that of the Q band.

Q. DIMERS

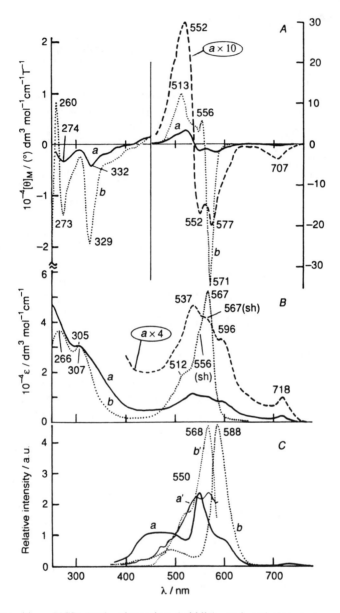

Figure 11   MCD (top), absorption (middle), and emission and excitation (bottom) spectra of **246** (M=BBr, curves a and a') and **242** (curves b and b') in deaerated chloroform.   Note the encircled magnification factor.

Concerning planar binuclear complexes, those of tetrabenzoporphy-rins (TBPs) **244** and **245** [249] and subphthalocyanines (SubPcs) **246** [246] are known, where two TBP or SubPc rings share a common benzene ring. The corresponding Pc dimers, **247-249**, were already reported [250]. ^{1}H NMR spectra are more complex compared with those of mononuclear control molecules because of the ring-current effect of two macrocycles [246, 249]. In all cases, the Q band broadens, and a weak peak appears to the longer-wavelength side of the Q band, suggesting that two included macrocycles are not completely flat in solution [251]. Also, the quantum yields of the S_1 emission of the dimers are about one-order of magnitude smaller than those of the corresponding monomers [246, 249, 250], suggesting effective intramolecular quenching. The potential difference between the first ring oxidation and reduction in **245** (1.63 V) was slightly smaller than that in the corresponding monomer (1.75 V), consistent with the longer wavelngth shift of the absorption and MCD spectra [249]. In **245**, the reduction and oxidation of two macrocycles occur stepwise at closely located potentials.

Although the structures are not shown here, cofacial dimers are divided into two groups. One is the sandwich-type dimers, which consist of two macrocycles and one metal of the lanthanum series, and the other is the μ-oxo complexes. Concerning sandwich-type dimers, those consisting of two Ncs with lutetium [57], two Acs with lutetium [252], two tetrapyrazinoporphyrazines with ytterbium [253], and a Nc and a Ac or a Pc and a Nc with lutetium [254] have been reported to date. Homodimers are obtained by fusing aromatic *ortho*-dinitriles in the presence of metal salts of lanthanum series at temperatures of less than ~290°C. In order to obtain heterodimers, two methods are known. In one method, lanthanum metal insertion reaction is carried out in a mixture of two Pc analogues, while in the other method, a Pc analogue containing one lanthanum metal is reacted with disodium or dilithium salt of another Pc analgue. In both cases, the desired heterodimer is separated from a reaction mixture by column chromatography.

Recent absorption and MCD studies [255] of Lu^{III}(Nc)(Pc) indicated that the hole is delocalized on both of the two macrocycles. The Q band of Lu^{III}(Nc)(Pc) (704 nm in chloroform) exists inbetween that of Lu^{III}(Pc)$_2$ (658 nm) and Lu^{III}(Nc)$_2$ (758 nm).

As a μ-oxo dimer, germanium tetrabenzotriazacorrole (GeTBC) is known [248]. This dimer is obtained as blue-black fine needles by heating GeTBC coordinated by a OH$^-$ at the fifth position at 500°C. Intense IR bands were detected at 892 and 867 cm^{-1} which are ascribable to Ge-O-Ge bond.

244   M=H₂
245   M=Cu

246   M=BF, BCl, or BBr

247   M=H₂
248   M=Zn
249   M=Co

R. CONCLUDING REMARKS

The synthesis and purification of Pc analogues generally becomes difficult with increasing size of the molecules, mainly because of the low stability and solubility of large molecules. Accordingly, the synthetic procedures for the larger molecules have been described in more detail than for the small molecules, because even slight changes in conditions can easily lead to an unsuccessful result. It is recommended that the literature be consulted when preparing Pc analogues. Almost no compound seems to be synthesized as easily as the parent Pcs.

ACKNOWLEDGMENTS

The author would like to express his sincere gratitude to Dr. M. Fukushima for his special work in preparing the structural figures, and to Dr. W. A. Nevin for his considerable help in preparing the text. Without their help, this manuscript would not have been accomplished. I would also like to thank Professor. E. A. Luk'yanets for his kind gift of Ref. [1], which has helped a great deal, especially with regard to the Russian literature.

REFERENCES

1.   E. A. Luk'yanets, *Electronic Spectra of Phthalocyanines and Related Compounds*, Tcherkassy, 1989.
2.   A. H. Cook and R. P. Linstead, *J. Chem. Soc.* (1937) 929.
3.   M. J. Camennzino and C. L. Hill, *Inorg. Chim. Acta*, 99 (1985) 63.
4.   N. Kobayashi et al., to be submitted.
5.   R. P. Linstead and M. Whalley, *J. Chem. Soc.* (1952) 4839.
6.   P. M. Brown, J. B. Speiers, and M. Whalley, *J. Chem. Soc.* (1957) 2882.
7.   M. E. Baguley, H. France, R. P. Linstead, and M. Whalley, *J. Chem. Soc.* (1955) 3521.
8.   C. E. Ficken and R. P. Linstead, *J. Chem. Soc.* (1952) 4846.
9.   R. P. Linstead, *J. Chem. Soc.* (1953) 2873.
10.   (a) V. N. Kopranenkov, L. S. Goncharova, and E. A. Luk'yanets, *J. Gen. Chem. USSR*, 47 (1977) 1954. (b) S. V. Barkanova, V. M. Derkacheva, I. A. Zheltukhin, O. L. Kaliya, V. N. Kopranenkov, and E. A. Luk'yanets, *ibid.*, 21 (1985) 1848.
11.   V. N. Kopranenkov, D. B. Askerov, A. M. Shul'ga, and E. A. Luk'yanets, *Chem. Heterocycl. Comp.* (1988) 1043.
12.   V. N. Kopranenkov and G. I. Rumyantseva, *J. Gen. Chem. USSR*, 45 (1975) 1521.
13.   L. E. Marinina, S. A. Mikhalenko, and E. A. Luk'yanets, *J. Gen. Chem. USSR*, 43 (1973) 2010.
14.   V. N. Kopranenkov, L. S. Goncharova, and E. A. Luk'yanets, *J. Gen. Chem. USSR*, 15 (1979) 962.
15.   V. N. Kopranenkov, L. S. Goncharova, L. E. Marinina and E. A. Luk'yanets, *Chem. Heterocycl. Comp.* (1982) 1269.

16.	V. N. Kopranenkov, L. S. Goncharova, L. E. Marinina, and E. A. Luk'yanets, *J. Gen. Chem. USSR*, 49 (1979) 1233.
17.	I. K. Shushkevich, V. N. Kopranenkov, S. S. Dvornikov, and K. N. Solovyov, *J. Appl. Spectrosc.*, 46 (1987) 368.
18.	W. Wolf, E. Degener, and S. Petersen, *Angew. Chem.*, 72 (1960) 963.
19.	G. P. Shaposhnikov, V. P. Kulinich, Yu. M. Osipov, and R. P. Smirnov, *Chem. Heterocycl. Comp.* (1986) 1036.
20.	C. J. Schramm and B. M. Hoffman, *Inorg. Chem.*, 19 (1980) 383.
21.	C. S. Velazquez, W. E. Broderick, M. Sabat, A. G. M. Barrett, and B. M. Hoffman, *J. Am. Chem. Soc.*, 112 (1990) 7408.
22.	M. Whalley, *J. Chem. Soc.* (1961) 866.
23.	M. J. Stillman and T. Nyokong, in *Phthalocyanines–Properties and Applications*, Eds. C. C. Leznoff and A. B. P. Lever, VCH, NY(1989), ch. 3.
24.	M. Gouterman, in *The Porphyrins*, Ed. D. Dolphin, Academic Press, NY, London (1978), vol. III, ch. 1.
25.	S. B. Piepho and P. N. Schats, *Group Theory in Spectroscopy–With Applications to MCD*, Wiley, New York, 1983.
26.	M. Gouterman, *J. Mol. Spectrosc.*, 6 (1961) 138.
27.	C. Weiss, H. Kobayashi, and M. Gouterman, *J. Mol. Spectrosc.*, 6 (1965) 415.
28.	H. Shimizu, A. Kaito, and M. Hatano, *Bull. Chem. Soc. Jpn.*, 52 (1979) 2678.
29.	L. M. Blinov, V. N. Kopranenkov, S. P. Palto, and S. G. Yudin, *Opt. Spectrosc.*, 62 (1987) 631.
30.	(a) J. W. Dodd and N. S. Hush, *J. Chem. Soc.* (1964) 4607. (b) D. W. Clack and N. S. Hush, *J. Am. Chem. Soc.*, 87 (1965) 4238. (c) N. S. Hush, *Theor. Chim. Acta*, 4 (1966) 108.
31.	G. E. Ficken, R. P. Linstead, E. Stephen, and M. Whalley, *J. Chem. Soc.* 1958) 3879.
32.	H. P. H. Thijssen and S. Volker, *Chem. Phys. Lett.*, 82 (1981) 478.
33.	S. M. Arabei, G. D. Egorova, K. N. Solovev, and S. F. Shkirman, *Opt. Spectrosc.*, 59 (1985) 296.
34.	M. Gouterman, G. H. Wagniere, and L. C. Snyder, *J. Mol. Spectrosc.*, 11 (1963) 108.
35.	(a) A. M. Schaffer and M. Gouterman, *Theor. Chim. Acta (Berl.)*, 25 (1972) 62. (b) A. M. Schaffer, M. Gouterman, and E. R. Davidson, *ibid.*, 30 (1973) 9. (c) A. J. McHugh, M. Gouterman, and C. Weiss, Jr., *ibid.*, 24 (1972) 346.
36.	S. S. Dvornikov, V. N. Knyukshto, V. A. Kuzmitsky, A. M. Shulga and K. N. Solovyov, *J. Luminescence*, 23 (1981) 373.
37.	L. K. Lee, N. H. Sabelli, and P. R. LeBreton, *J. Phys. Chem.*, 86 (1982) 3926.
38.	Z. Berkovitch-Yellin and D. E. Ellis, *J. Am. Chem. Soc.*, 103 (1981) 6066.
39.	G. P. Gurinovich, G. N. Sinyakov, and A. M. Shulga, *Izv. Akad. Nauk SSSR, Ser. fiz.*, 34 (1970) 620.
40.	C. G. Barraclough, R. L. Martin, S. Mitra, and R. C. Sherwood, *J. Chem. Phys.*, 53 (1970) 1638.
41.	B. Gonzalez, J. Kouba, S. Yee, C. A. Reed, J. F. Kirner, and W. R. Scheidt, *J. Am. Chem. Soc.*, 97 (1975) 3247.
42.	O. G. Khelevina, P. A. Stuzhin, and B. D. Berezin, *Chem. Heterocycl. Comp.* (1984) 757.
43.	P. Doppelt and S. Huille, *New J. Chem.*, 14 (1990) 607.
44.	E. F. Bradbrook and R. P. Linstead, *J. Chem. Soc.* (1936) 1744.
45.	Yu. M. Gryaznov, O. L. Lebedev, and A. A. Chastov, *Opt. Spectrosc.*, 20 (1966) 278.
46.	M. Hanack, G. Renz, J. Strahle, and S. Schmid, *Chem. Ber.*, 121 (1988) 1479; *idem. J. Org. Chem.*, 56 (1991) 3501.
47.	J. Metz, O. Schneider and M. Hanack, *Inorg. Chem.*, 23 (1984) 1065.
48.	G. A. Yurlova, *J. Gen. Chem. USSR.* 41 (1971) 1333.

49. T. Ohya, J. Tanaka, N. Kobayashi, and M. Sato, *Inorg. Chem.*, 29 (1990) 3734.
50. S. A. Mikhalenko and E. A. Luk'yanets, *J. Gen. Chem. USSR*, 39 (1969) 2495.
51. A. Vogler and H. Kunkely, *Inorg. Chim. Acta*, 44 (1980) L209.
52. S. Deger and H. Hanack, *Synth. Met.*, 13 (1986) 319; V. Keppeler, S. Degar, A. Lange, and M. Hanack, *Angew. Chem. Int. Ed. Engl.*, 26 (1987) 344.
53. S. Deger and M. Hanack, *Isr. J. Chem.*, 27 (1986) 347.
54. E. I. Kovshev and E. A. Luk'yanets, *J. Gen. Chem. USSR*, 42 (1972) 691.
55. K. Ito, H. Ueda, N. Kodera, S. Kosuke, H. Kuruma, and K. Tanimoto, *Jpn. Patent*; *Chem. Abstr.* 71 (1969) 70299u.
56. E. I. Kovshev, V. A. Puchnova, and E. A. Luk'yanets, *J. Org. Chem. USSR*, 7 (1971) 364.
57. M. G. Gal'pern, T. D. Talismanova, L. G. Tomilova, and E. A. Luk'yanets, *J. Gen. Chem. USSR*, 55 (1985) 980.
58. M. G. Gar'pern, E. G. Dikolenko, V. F. Donyagina, S. D. Isaev, O. F. Kozlov, and E. A. Luk'yanets, *Zh. Obshch. Khim.*, 50 (1980) 2390.
59. T. A. Shatskaya, M. G. Gal'pern, V. R. Skvarchenko, and E. A. Luk'yanets, *J. Gen. Chem. USSR*, 57 (1987) 2115.
60. M. G. Gal'pern, V. K. Shalaev, T. A. Shatsskaya, L. S. Shiskanova, V. R. Skvarchenko, and E. A. Luk'yanets, *J. Gen. Chem. USSR*, 53 (1983) 2346.
61. G. I. Gonsharova, M. G. Gal'pern, and E. A. Luk'yanets, *J. Gen. Chem. USSR*, 52 (1982) 581.
62. A. Michael and J. E. Bucher, *Am. Chem. J.*, 20 (1898) 89.
63. E. I. Kovshev and E. A. Luk'yanets, *J. Gen. Chem. USSR*, 42 (1972) 1584.
64. W. Freyer and L. Q. Minh, *J. Prakt. Chem.*, 329 (1987) 365.
65. M. J. Cook, A. J. Dunn, S. D. Howe, and A. J. Thomson, *J. Chem. Soc., Perkin Trans. I*, (1988) 2453.
66. W. E. Ford, B. D. Richter, M. E. Kenney, and M. A. Rodgers, *Photochem. Photobiol.*, 50 (1989) 277.
67. Y. Iwakabe, S. Numata, N. Kinjo, and A. Kakuta, *Bull. Chem. Soc. Jpn.*, 63 (1990) 2734.
68. B. L. Wheeler, G. Nagasubramanian, A. J. Bard, L. A. Schechtman, D. R. Dininny, and M. E. Kenney, *J. Am. Chem. Soc.*, 106 (1984) 7404.
69. Y. Ikeda, Master's Thesis, Tohoku University, 1989.
70. M. L. Kaplan, A. J. Lovinger, W. D. Reents, Jr., and P. H. Schmidt, *Mol. Cryst. Liq. Cryst.*, 112 (1984) 345.
71. C. G. Cannon and G. B. B. M. Sutherland, *Spectrochim. Acta*, 4 (1951) 373.
72. S. Hayashida and N. Hayashi, *Chem. Lett.* (1990) 2137.
73. W. Freyer and L. Q. Minh, *Monatsh. Chem.*, 117 (1986) 475.
74. W. Freyer, *Z. Chem.*, 26 (1986) 216, 217.
75. W. Freyer, L. Q. Minh, and K. Teuchner, *Z. Chem.*, 28 (1988) 25.
76. W. Freyer and K. Teuchner, *J. Photochem. Photobiol. A*, 45 (1988) 117.
77. P. A. Firey and M. A. J. Rodgers, *Photochem. Photobiol.*, 45 (1987) 535.
78. H. Yanagi, M. Ashida, J. Elbe, and D. Wöhrle, *J. Phys. Chem.*, 94 (1990) 7056.
79. A. Volker, H. -J. Adick, R. Schmidt, and H. -D. Brauer, *Chem. Phys. Lett.*, 159 (1989) 103.
80. (a) S. S. Iodoko, O. L. Kaliya, M. G. Gal'pern, V. N. Kopranenkov, O. L. Lebedev, and E. A. Luk'yanets, *Sov. J. Coord. Chem.*, 8 (1982) 552. (b) E. Orti, M. C. Piqueras, R. Crespo, and J. L. Bredas, *Chem. Mat.*, 2 (1990) 110.
81. S. Gaspard, M. Verdaguir, and R. Viouy, *J. Chim. Phys. Physcochim. Biol.*, 69 (1972) 1740; *idem.*, *J. Chem. Res. (S)* (1979) 271, and Refs. 3-7 cited therein.
82. D. L. Ledson and M. V. Twigg, *Inorg. Chim. Acta*, 13 (1975) 43.
83. N. el Khatib, B. Boudiema, M. Maitrot, H. Chermette, and L. Porte, *Can. J. Chem.*, 66 (1988) 2313.
84. W. E. Ford, B. D. Richter, M. A. J. Rogers, and M. E. Kenney, *J. Am. Chem. Soc.*, 111 (1989) 2362.

85. P. A. Firey, W. E. Ford, J. R. Sounik, M. E. Kenney, and M. A. J. Rodgers, *J. Am. Chem. Soc.*, 110 (1988) 7626.
86. V. M. Mokshin, O. A. Postnikova, M. G. Gal'pern, G. I. Goncharova, S. A. Mikhalenko, and E. A. Luk'yanets, *Sov. J. Chem. Phys.*, 4 (1987) 525.
87. G. Magner, M. Savy, G. Scarbeck, J. Riga, and J. J. Verbist, *J. Electrochem. Soc.*, 128 (1981) 1674; J. Riga, M. Savy, J .J. Verbist, E. Guerchais, and J. Sala-Pala, *J. Chem. Soc., Faraday Trans. 1*, 78 (1982) 2773.
88. (a) F. Coowar, O. Coutamin, M. Savy, and G. Scarbeck, *J. Electroanal. Chem.*, 246 (1988) 119. (b) D. Schlettwein, M. Kaneko, A. Yamada, D. Wöhre, and N. I. Jaeger, *J. Phys. Chem.*, 95 (1991) 1748.
89. J. R. Darwent, I. McCubbin, and G. Porter, *J. Chem. Soc., Faraday Trans. 2*, 78 (1982) 903.
90. I. McCubbin and D. Phillips, *J. Photochem.*, 34 (1986) 187.
91. N. C. Yates, J. Moan, and A. Western, *Photochem. Photobiol. B*, 4 (1990) 379.
92. V. N. Kopranenkov and E. A. Luk'yanets, *J. Gen. Chem. USSR.*, 41 (1971) 2366.
93. J. Rigaudy and M. M. Ricard, *Tetrahedron*, 24 (1968) 3241.
94. V. N. Kopranenkov,. G. I. Rumyantzeva, and E. A. Luk'yanets, *Anilino-Dyes Ind.* (1974) 6.
95. P. J. Brach, S. J. Grammatica, O. A. Ossanna and L. Weinberger, *J. Heterocycl. Chem.*, 7 (1970) 1403.
96. R. P. Linstead, E. G. Noble, and J. M. Wright, *J. Chem. Soc.* (1937) 911; J. S. Anderson, E. F. Bradbrook, A. H. Cook, and R. P. Linstead, *ibid.* (1938) 1151.
97. D. Wöhrle, J. Gitzel, I. Okura, and S. Aono, *J. Chem. Soc., Perkin Trans. 2*, (1985) 1171.
98. J. E. Scott, *Histcohem.*, 30 (1972) 215.
99. M. G. Gal'pern and E. A. Luk'yanets, *J. Gen. Chem. USSR*, 39 (1969) 2477.
100. C. Enger, *Chem. Ber.*, 27 (1894) 1784.
101. M. J. Danzig, C. Y. Liang, and E. Passaglia, *J. Am. Chem. Soc.*, 85 (1963) 668.
102. (a) B. Vanvlieberge, M. Z. Yang, F. X. Sauvage, M. G. Backer, and A. Chapput, *Spectrochim. Acta*, 42A (1986) 1133. (b) M. Hanack and R. Thies, *Chem. Ber.*, 121 (1988) 1225.
103. S. Iwashima and T. Sawada, *Res. Bull. Meisei Univ. Phys. Sci. Eng.*, 19 (1983) 43; *Chem. Abstr.*, 99 (1983) 141522.
104. M. Yokote, F. Shibamiya, and N. Sakikubo, *Yuki Gosei Kagaku Kyokai Shi*, 27 (1969) 448; *Chem. Abstr.*, 71 (1969) 125956p. M. Yokote and F. Shibamiya, *Kogyo Kagaku Zasshi*, 61 (1958) 994. M. Yokote, F. Shibamiya, and H. Yokomizo, *Yuki Gousei Kyokai Shi*, 27 (1969) 340. M. Yokote, F. Shibamiya, and N. Sakikubo, *ibid.*, 27 (1969) 448.
105. F. Fukada, *Nippon Kagaku Zasshi*, 78 (1957) 1348; *Chem. Abstr.*, 53 (1957) 21339b.
106. T. D. Smith, J. Livorness, and H. Taylor, *J. Chem. Soc., Dalton Trans.* (1983) 1391.
107. A. Skorobogaty, T. D. Smith, G. Dougherty, and J. P. Pilbrow, *J. Chem. Soc., Dalton Trans.* (1985) 651.
108. K. Kasuga, M. Morisada, M. Handa, and K. Sogabe, *Inorg. Chim. Acta*, 174 (1990) 161.
109. J. E. Scott, *Histochem.*, 32 (1972) 191.
110. M. Yokote, F. Shibamiya, and S. Shoji, *Kogyo Kagaku Zasshi*, 67 (1964) 166.
111. E. G. Gal'pern and E. A. Luk'yanets, *Acad. Nauk SSSR. Bull. Chem. Sci.*, 22 (1973) 1925.
112. S. Palacin, A. Ruaudel-Teixier, and A. Barraud, *J. Phys. Chem.*, 90 (1986) 6237.
113. S. Palacin, A. Ruaudel-Teixier, and A. Barraud, *J. Phys. Chem.*, 93 (1989) 7195; S. Palacin, *Thin Solid Films*, 178 (1989) 327.

114. A. S. Akopov, B. D. Berezin, and V. V. Bykova, *Theo. Expl. Chem.*, 15 (1979) 357.
115. (a) C. Fierro, A. B. Anderson, and D. A. Scherson, *J. Phys. Chem.*, 92 (1988) 6902. (b) V. A. Kuz'mitskii, K. N. Solov'ev, and V. N. Kopranenkov, *J. Appl. Spectrosc.*, (1989) 1165.
116. S. Palacin and A. Barraud, *J. Chem. Soc., Chem. Commun.* (1989) 45.
117. A. Skorobogaty, T. D. Smith, J. R. Pilbrow, and S. J. Rawlings, *J. Chem. Soc., Faraday Trans. 1*, 82 (1986) 173.
118. M. Z. Yang, M. G. De Backer, and F. X. Sauvage, *New J. Chem.*, 4 (1990) 273.
119. H. Kropf and H. Hoffmann, *Tet. Lett.* (1967) 659.
120. K. Sakamoto, J. Sonobe, and F. Shibamiya, *Nihon Kagaku Kaishi* (1990) 770.
121. H. W. Rothkopf, D. Wöhrle, R. Muller, and G. Kossmehl, *Chem. Ber.*, 108 (1975) 875.
122. N. Kobayashi, K. Adachi, and T. Osa, *Anal. Sci.*, 6 (1990) 449.
123. M. G. Gal'pern and E. A. Luk'yanets, *Khim. Geterotsikl. Soed.* (1972) 858; *Chem. Abstr.*, 77 (1972) 75197t.
124. S. Tokita, M. Kojima, N. Kai, K. Kurogi, H. Nishi, H. Tomoda, S. Saito, and S. Shiraishi, *Hihon Kagaku Kaishi* (1990) 219.
125. K. Ohta, T. Watanabe, T. Fujimoto, and I. Yamamoto, *J. Chem. Soc., Chem. Commun.* (1989) 1611.
126. S. Greenberg, S. M. Marcuccio, C. C. Leznoff, and K. B. Tomer, *Synthesis* (1986) 406; M. Hanack and G. Pawlowski, *ibid.* (1980) 287; C. C. Leznoff, S. M. Marcuccio, S. Greenberg, and A. B. P. Lever, *Can. J. Chem.*, 63 (1985) 623.
127. A. B. P. Lever, S. Licoccia, K. Magnell, P. C. Minor, and B. S. Ramaswamy, *Adv. Chem. Ser.*, 201 (1982) 237.
128. V. E. Maizlish, A. B. Korzhenevskii, and V. N. Klyuev, *Chem. Heterocycl. Comp.* (1984) 1031.
129. M. G. Gal'pern and E. A. Luk'yanets, *Zh. Vses. Khim. Obsch.*, 12 (1967) 74; *Chem. Abstr.*, 68 (1968) 2789s.
130. M. G. Gal'pern and E. A. Luk'yanets, *J. Gen. Chem. USSR*, 41 (1971) 2579.
131. Ref. 8 cited in F. Beck, *Ber. Bunsenges. Phys. Chem.*, 77 (1973) 353.
132. J. A. Elvidge and R. P. Linstead, *J. Chem. Soc.* (1955) 3536.
133. J. A. Elvidge, J. H. Golden, and R. P. Linstead, *J. Chem. Soc.* (1957) 2466.
134. N. Kobayashi, R. Kondo, S. Nakajima, and T. Osa, *J. Am. Chem. Soc.*, 112 (1990) 9640.
134a. C. C. Leznoff, Y. Qin, and C. R. McArther, submitted for publication.
135. J. H. Helberger, A. von Rebay, and D. B. Hever, *Justus Liebigs Ann. Chem.*, 533 (1938) 197.
136. J. H. Helberger and D. B. Hever, *Justus Liebigs Ann. Chem.*, 536 (1938) 173.
137. P. A. Barrett, R. P. Linstead, F. G. Rundal, and G. A. P. Tuey, *J. Chem. Soc.* (1940) 1079.
138. R. P. Linstead and F. T. Weiss, *J. Chem. Soc.* (1950) 2975.
139. A. Vogler and H. Kunkely, *Angew. Chem. Int. Ed. Engl.*, 17 (1978) 760.
140. H. L. Yale, *J. Am. Chem. Soc.*, 69 (1947) 1547.
141. V. N. Kopranenkov, E. A. Makarova, and E. A. Luk'yanets, *J. Gen. Chem. USSR*, 51 (1981) 2353.
142. D. E. Remy, *Tet. Lett.* 24 (1983) 1451.
143. V. N. Kopranenkov, E. A. Makarova, and S. N. Dashkevich, *Chem. Heterocycl. Comp.* (1985) 1126; *ibid.* (1982) 1563.
144. R. Bonnett and R. F. C. Brown, *J. Chem. Soc., Chem. Commun.* (1972) 393.
145. M. A. Kolesnikov, N. A. Red'kin, and A. I. Tochilkin, *Byull. Izobret.*, No. 42 (1971) 30.
146. A. Vogler, H. Kunkely, and B. Rethwisch, *Inorg. Chim. Acta*, 46 (1980) 101.
147. R. B. M. Koehorst, J. F. K. Kleibenker, T. J. Schaafsma, D. A. B. B. Geurtsen, R. N. Henrie, and H. C. Plas, *J. Chem. Soc., Perkin Trans. 2* (1981) 1005.

148. M. Hanack amd T. Zipplies, *J. Am. Chem. Soc.*, 107 (1985) 6127.
149. J. Martinsen, L. J. Pace, T. E. Phillips, B. M. Hoffman, and J. A. Ibers, *J. Am. Chem. Soc.*, 104 (1982) 83.
150. N. Kobayashi, M. Koshiyama, and T. Osa, *Inorg. Chem.*, 24 (1985) 2502; *idem.*, *Chem. Lett.* (1983) 163.
151. K. Fischer and M. Hanack, *Angew. Chem. Int. Ed. Engl.*, 22 (1983) 724.
152. (a) C. O. Bender, R. Bonnett, and R. G. Smith, *J. Chem. Soc. C*, (1970) 1251. (b) *idem.*, *J. Chem. Soc., Perkin Trans. 1* (1972) 771. (c) *idem.*, *J. Chem. Soc., Chem. Commun.* (1969) 345.
153. D. Dolphin, J. R. Sams, and T. B. Tsin, *Inorg. Synth.*, 20 (1980) 155.
154. T. E. Phillips and B. M. Hoffman, *J. Am. Chem. Soc.*, 99 (1977) 7734.
155. C. O. Bender and R. Bonnett, *J. Chem. Soc., Chem. Commun.* (1966) 198.
156. J. Bornstein, D. E. Remy, and J. E. Shields, *Tet. Lett.* (1974) 4247.
157. V. N. Kopranenkov, T. A. Tarkhanova, and E. A. Luk'yanets, *J. Org. Chem. USSR*, (1979) 570.
158. V. N. Kopranenkov, E. A. Makarova, S. N. Dashkevich, and E. A. Luk'yanets, *Chem. Heterocycl. Comp.* (1988) 630.
159. V. N. Kopranenkov, S. N. Dashkevich, V. K. Shevtsov, and E. A. Luk'yanets, *Chem. Heterocycl. Comp.* (1983) 52.
160. V. N. Kopranenkov, S. N. Dashkevich, and E. A. Luk'yanets, *J. Gen. Chem. USSR*, 51 (1981) 2165.
161. S. N. Dashkevich and E. A. Luk'yanets, unpublished data.
162. V. N. Kopranenkov, E. A. Makarova, and E. A. Luk'yanets, *Chem. Heterocycl. Comp.* (1986) 960.
163. N. Kobayashi, M. Numao, and T. Osa, to be submitted.
164. (a) K. Ichimura, M. Sakuragi, H. Morii, M. Yasuike, M. Fukui, and O. Ohno, *Inorg. Chim. Acta*, 176 (1990) 31. (b) *idem., ibid.* 182 (1991) 83; 186 (1991) 95.
165. T. E. Phillips, R. P. Scaringe, B. M. Hoffman, and J. A. Ibers, *J. Am. Chem. Soc.*, 102 (1980) 3435.
166. J. M. Robertson, *J. Chem. Soc.* (1936) 1195. R. P. Linstead and J. M. Robertson, *ibid.* (1936) 1736.
167. L. Edwards, M. Gouterman, and C. B. Rose, *J. Am. Chem. Soc.*, 98 (1976) 7638.
168. K. N. Solov'ev, V. A. Mashenkov, A. T. Gradyushko, A. E. Turkova, and V. P. Lezina, *Zh. Prikl. Spectrosk.*, 13 (1976) 339.
169. A. N. Sevchenko, K. N. Solov'ev, S. F. Shkirman, and T. F. Kachura, *Sov. Phys. Dokl.*, 10 (1965) 349.
170. A. N. Sevchenko, K. N. Solov'ev, A. T. Gradyushko, and S. F. Shkirman, *Dokl. Akad. Nauk SSSR*, 169 (1966) 77; *Chem. Abstr.*, 65 (1966) 14668b.
171. A. Vogler, B. Rethwisch, H. Kunkely, and J. Huttermann, *Angew. Chem. Int. Ed. Engl.*, 17 (1978) 951.
172. (a) J. C. Goedheer and J. P. J. Siero, *Photochem. Photobiol.*, 6 (1967) 509. (b) J. C. Goedheer, *ibid.*, 6 (1967) 521.
173. T. J. Aartsma, M. Gouterman, C. Jochum, A. L. Kwiram, B. V. Pepich, and L. D. Williams, *J. Am. Chem. Soc.*, 104 (1982) 6278.
174. B. R. James, K. J. Reimer, and T. C. T. Wong, *J. Am. Chem. Soc.*, 99 (1977) 4815.
175. K. J. Reimer, M. M. Reimer, and M. J. Stillman, *Can. J. Chem.*, 59 (1981) 1388.
176. P. E. Fielding and A. W. H. Mau, *Aust. J. Chem.*, 29 (1976) 933.
177. R. J. Platenkamp and G. W. Canters, *J. Phys. Chem.*, 85 (1981) 56.
178. A. Vogler, B. Rethwisch, H. Kunkely, and H. Huttermann, *Angew. Chem. Int. Ed. Engl.*, 17 (1978) 952.
179. T. Yamamoto, T. Nozawa, N. Kobayashi, and M. Hatano, *Bull. Chem. Soc. Jpn.*, 55 (1982) 3059.
180. N. Kobayashi and T. Osa, *Chem. Pharm. Bull.*, 37 (1989) 3105. N. Kobayashi, *Inorg. Chem.*, 24 (1985) 3324.

181. A. N. Sevchenko, K. N. Solov'ev, A. T. Gradyushko, and S. F. Shkirman, *Sov. Phys. Dokl.*, 11 (1967) 587.
182. K. N. Solov'ev, S. F. Shkirman, and T. F. Kachura, *Akad. Nauk SSSR. Bull. Phys. Ser.*, 27 (1963) 763.
183. I. E. Zalesskii, V. N. Kotlo, A. N. Sevchenko, K. N. Solov'ev, and S. F. Shkirman, *Sov. Phys. Dokl.*, 18 (1973) 320; 19 (1975) 589.
184. G. P. Gradyushko, A. N. Sevchenko, K. N. Solov'ev, and M. P. Tsvirko, *Photochem. Photobiol.*, 11 (1970) 387.
185. K. N. Solov'ev, M. P. Tsvirko, and T. F. Kachura, *Opt. Spectrosc.*, 40 (1976) 391.
186. G. F. Stelmakh and M. P. Tsvirko, *Opt. Spectrosc.*, 55 (1983) 516.
187. J. Aaviksoo, A. Freiberg, S. Savikhin, G. F. Stelmakh, and M. P. Tsvirko, *Chem. Phys. Lett.*, 111 (1984) 275.
188. V. G. Maslov, *Opt. Spectrosc.*, 50 (1981) 599.
189. V. A. Lyubimtsev, *Opt. Spectrosc.*, 55 (1983) 624.
190. L. Bajema, M. Gouterman, and C. B. Rose, *J. Mol. Spectrosc.*, 39 (1971) 421.
191. A. T. Gradyushko and M. P. Tsvirko, *Opt. Spectrosc.*, 31 (1971) 291.
192. S. M. Arabei, K. N. Solov'ev, S. F. Shkirman, and T. F. Kachura, *Zh. Prikl. Spektrosk.*, 50 (1989) 954; *Chem. Abstr.*, 111 (1989) 6596u.
193. H. D. Forsterling and H. Kuhn, *Int. J. Quant. Chem.*, 2 (1968) 413.
194. K. N. Solov'ev, V. A. Mashenkov, and T. F. Kachura, *Opt. Spectrosc.*, 27 (1969) 24.
195. W. H. Henneker, M. Pawlikowski, S. Siebrand, and M. Z. Zgierski, *J. Phys. Chem.*, 87 (1983) 4805.
196. L. K. Hanson, C. K. Chang, M. S. Davis, and J. Fajer, *J. Am. Chem. Soc.*, 103 (1981) 663.
197. H. S. Hush and J. R. Rowlands, *J. Am. Chem. Soc.*, 89 (1967) 2976.
198. P. W. Lau and W. C. Lin, *J. Inorg. Nucl. Chem.*, 37 (1975) 2389.
199. W. C. Lin, *Inorg. Chem.*, 15 (1976) 1114.
200. C. P. S. Taylor, *Biochim. Biophys. Acta*, 491 (1977) 137; J. S. Griffith, *The Theory of Transition-Metal Ions*, Cambridge University Press, New York, 1967; J. S. Griffith, *Nature (London)*, 180 (1957) 30.
201. J. R. Sams and T. B. Tsin, *Chem. Phys. Lett.*, 25 (1974) 599.
202. A. B. Smirnov and V. Khleskov, *Teor. Eksp. Khim.*, 25 (1989) 601; *Chem. Abstr.*, 112 (1990) 65930d.
203. K. J. Reimer, C. A. Sibley, and J. R. Sams, *J. Am. Chem. Soc.*, 105 (1983) 5147.
204. K. J. Reimer and M. M. Reimer, *Inorg. Chim. Acta*, 56 (1981) L5.
205. J. -H. Fuhrhop, in *Porphyrins and Metalloporphyrins*, Ed. K. M. Smith, Elsevier, Amsterdam, 1975, ch. 14.
206. C. H. Rein, M. Hanack, K. Peters, E. M. Peters, and H. G. Schnering, *Inorg. Chem.*, 26 (1987) 2647.
207. J. A. R. van Veen, J. F. van Baar, C. J. Croese, J. G. F. Coolegem, N. D. Wit, and H. A. Colijn, *Ber. Bunsenges. Phys. Chem.*, 85 (1981) 693.
208. V. S. Bagotskii, M. R. Tarasevich, K. A. Radyushkina, O. A. Levina, and S. I. Andruseva, *J. Power Sources*, 2 (1978) 233.
209. T. P. Carter, C. Braeuchle, V. Y. Lee, M. Manavi, and W. E. Moerner, *J. Phys. Chem.*, 91 (1987) 3998.
210. L. Kador, D. Haarer, and R. I. Personov, *J. Chem. Phys.*, 86 (1987) 5300.
211. K. Yamashita, Y. Harima, H. Kubota, and H. Suzuki, *Bull. Chem. Soc. Jpn.*, 60 (1987) 803.
212. D. Wöhrle, in *Phthalocyanines-Properties and Applications*, Eds. C. C. Leznoff and A. B. P. Lever, VCH, NY (1989), ch. 2.
213. C. L. Honeybourne, *J. Chem. Soc., Chem. Commun.* (1982) 744; *Mol. Phys.*, 50 (1983) 1045. C. L. Honeybourne and R. J. Ewen, *J. Phys. Chem. Solids*, 45 (1984) 433.

214.   K. Yakushi, H. Yamakado, M. Yoshitake, N. Kosugi, H. Kuroda, A. Kawamoto, J. Tanaka, T. Sugano, M. Kinoshita, and S. Hino, *Syn. Metals*, 29 (1989) F95.
215.   T. Ida, H. Yamamoto, H. Masuda, and K. Yakushi, *Mol. Cryst. Liq. Cryst.*, 181 (1990) 243.
216.   K. Yakushi, M. Yoshitake, H. Kuroda, A. Kawamoto, J. Tanaka, T. Sugano, and M. Kinoshita, *Bull. Chem. Soc. Jpn.*, 61 (1988) 1571.
217.   M. Hanack, A. Lange, M. Rein, R. Rehnisch, G. Renz, and A. Leverenttz, *Syn. Metals*, 29 (1989) F1.
218.   V. N. Kopranenkov, A. M. Vorotnikov, S. N. Dashkevich, and E. A. Luk'yanets, *J. Org. Chem. USSR*, (1985) 803.
219.   V. N. Kopranenkov, A. M. Vorotnikov, and E. A. Luk'yanets, *J. Org. Chem. USSR*, (1980) 2467.
220.   M. Rein and M. Hanack, *Chem. Ber.*, 121 (1988) 1601.
221.   B. A. Goodman and J. B. Raynor, *Adv. Inorg. Chem. Radiochem.*, 13 (1970) 135.
222.   V. A. Kuz'mitskii, K. N. Solov'ev, and V. N. Kopranenkov, *J. Appl. Spectrosc.*, (1989) 1037.
223.   V. N. Kopranenkov, A. N. Vorotnikov, T. M. Ivanova, and E. A. Luk'yanets, *Chem. Heterocycl. Comp.* (1988) 1120.
224.   T. Ashida, Bachelor's Thesis, Tohoku University, 1990.
225.   N. Kobayashi, T. Ashida, and T. Osa, in preparation.
226.   J. H. Helberger, *Justus Liebigs Ann. Chem.*, 529 (1937) 205.
227.   T. F. Kachura, V. A. Mashenkov, K. N. Solov'ev, and S. F. Shkirman, *Vesti Akad. Nauk Beloruss. SSR., Ser. Khim. Nav.* (1969) 65; *Chem. Abstr.*, 71 (1969) 49914f.
228.   E. A. Makarova, V. N. Kopranenkov, V. K. Shevtsov, and E. A. Luk'yanets, *Khim. Geterotsikl. Soedin.* (1989) 1385; *Chem. Abstr.*,113 (1990) 23475z.
229.   P. A. Barrett, R. P. Linstead, and G. A. P. Tuey, *J. Chem. Soc.* (1939) 1809.
230.   J. H. Helberger and A. Rebay, *Justus Liebigs Ann. Chem.*, 531 (1937) 279.
231.   C. C. Leznoff and N. B. McKeown, *J. Org. Chem.*, 55 (1990) 2186.
232.   I. Woodward, *J. Chem. Soc.* (1940) 1.
233.   I. M. Das and B. Chaudhuri, *Acta Cryst.*, B28 (1972) 579.
234.   J. M. Robertson, *J. Chem. Soc.* (1939) 1811.
235.   N. Yu. Borovkov and A. S. Akopov, *Zh. Strukt. Khim.*, 28 (1987) 175; *Chem. Abstr.*, 107 (1987) 236325p.
236.   (a) V. N. Kolto, K. N. Solov'ev, and S. F. Shkirman, *Izv. Akad. Nauk SSSR, Ser. Fiz.*, 39 (1975) 1972; *Chem. Abstr.*, 84 (1976) 23911w. (b) A. N. Sevchenko, V. A. Mashenkov, and K. N. Solov'ev, *Dokl. Acad. Nauk SSSR*, 179 (1968) 61; *Chem. Abstr.*, 69 (1968) 14559a. (c) D. A. Savel'ev, I. P. Kotlyar, and A. N. Sidrov, *Zh. Fiz. Khim.*, 43 (1967) 1914; *Chem. Abstr.*, 72 (1970) 11824j.
237.   W. B. Euler, J. Martinsen, L. J. Pace, B. M. Hoffman, and J. A. Ibers, *Mol. Cryst. Liq. Cryst.*, 81 (1982) 231.
238.   K. Liou, M. Y. Ogawa, T. P. Newcomb, G. Quirion, M. Lee, M. Poirier, W. P. Halperin, B. M. Hoffman, and J. A. Ibers, *Inorg. Chem.*, 28 (1989) 3889.
239.   G. Quirion, M. Poirier, K. K. Liou, M. Y. Ogawa, and B. M. Hoffman, *Phys. Rev. B*, 37 (1988) 4272; *Solid State Commun.*, 64 (1987) 613.
240.   J. E. Bloor, J. Schlabitz, C. C. Walden, and A. Demerdache, *Can. J. Chem.*, 42 (1964) 2201.
241.   V. W. Day, T. J. Marks, and W. A. Wachter, *J. Am. Chem. Soc.*, 97 (1975) 4519.
242.   T. J. Marks and D. R. Stojakovic, *J. Am. Chem. Soc.*, 100 (1978) 1695.
243.   E. A. Cuellar and T. J. Marks, *Inorg. Chem.*, 20 (1981) 3766.
244.   A. B. P. Lever, *Adv. Inorg. Chem. Radiochem.*, 7 (1965) 27.
245.   A. Meller and A. Ossko, *Monash. Chem.*, 103 (1972) 150.
246.   N. Kobayashi, *J. Chem. Soc., Chem. Commun.* (1991) 1203.
247.   H. Kietaibl, *Monash. Chem.*, 105 (1974) 405.

248. M. Hujiki, H. Tabei, and K. Isa, *J. Am. Chem. Soc.*, 108 (1986) 1532.
249. N. Kobayashi, M. Numao, R. Kondo, S. -I. Nakajima, and T. Osa, *Inorg. Chem.* 30 (1991) 2241.
250. C. C. Leznoff, H. Lam, S. M. Marcuccio, W. A. Nevin, P. Janda, N. Kobayashi, and A. B. P. Lever, *J. Chem. Soc., Chem. Commu*n. (1987) 699.
251 M. Gouterman, D. Holten, and E. Lieberman, *Chem. Phys.*, 25 (1977) 139.
252. Y. Ishibashi, Bachelor Thesis, Tohoku University (1990).
253. K. Kasuga, K. Nishikoshi, T. Mihara, M. Handa, K. Sogabe, and K. Isa, *Inorg. Chem. Acta*, 174 (1990) 153.
254. N. B. Subbotin, L. G. Tomolova, E. V. Chernykh, N. A. Kostromina, and E. A. Luk'yanets, *J. Gen. Chem. USSR*, 56 (1986) 208.
255 N. Ishikawa, O. Ohno, and Y. Kaizu, *Chem. Phys. Lett.* 180 (1991) 51.

4

Thin Film

Phthalocyanine

Chemistry and

Technology

Tetsuo Saji

A. INTRODUCTION

The properties and applications of phthalocyanine and its metallo derivatives have been studied extensively [1-3]. These are due to their properties of high strength, chemical and light resistance, cleanliness, and durability. In addition, the function of these compounds is to serve as catalysts [4], electrochromic materials [5-7], sensitizers [8], photovoltaic materials [9, 10], color filters, and so on. In order to apply these compounds to devices, we have to prepare films of them. The technique for the preparation of phthalocyanine thin films is limited to vacuum sublimation and the LB film transfer technique due to their low solubility in usual organic solvents. The former technique is used for most unsubstituted phthalocyanine compounds. However, obtaining reproducible films is sometime difficult [11-15]. In order to control the morphology of the films, strict attention to sublimation conditions are required. The latter technique may be used for phthalocyanine compounds, which are substituted with long alkyl chains to make them soluble in volatile organic solvents [16, 17].

Recently, we have presented a novel technique for electrochemical formation of organic thin films by disruption of micellar aggregates by cationic surfactants having a ferrocenyl moiety [micelle disruption method (MD method)] [18-22]. Furthermore, we have presented the preparation of phthalocyanine thin films by this method using nonionic surfactant with ferrocenyl moiety (FPEG) (Fig. 1) [23-25]. Later studies showed that this phthalocyanine disperses by the surfactant and is not incorporated into the micelles. The pigment particles are released when the surfactants adsorbed on the particles are electrochemically oxidized and finally the particles deposit on the electrode (Fig. 2).

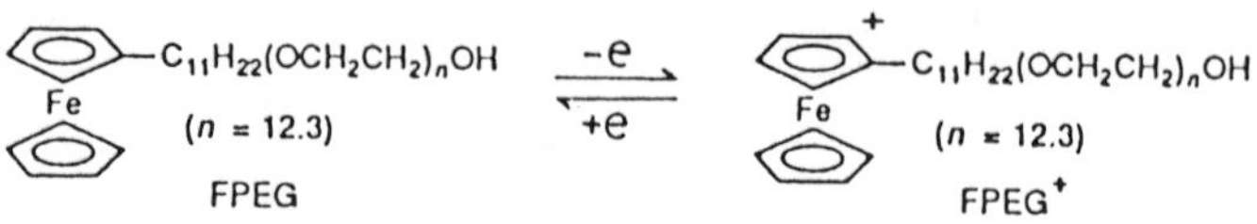

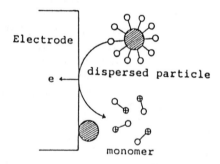

Figure 1 Molecular structures of nonionic surfactant with ferrocenyl moiety (FPEG) and its oxidant (FPEG$^+$).

Figure 2 Mechanism of formation of pigment thin film by electrolysis of surfactant with a ferrocenyl moiety.

The details of preparation and characterization of phthalocyanine thin films by the MD method are presented in the following sections.

B. PREPARATION AND PROPERTIES OF SURFACTANT

Surfactants have the properties to adsorb onto a solid particle and to produce by this adsorption energy barriers of sufficient height to disperse the particle in an aqueous solution. For the formation of electrical barriers to aggregation, ionic surfactants are generally used. For the formation of nonelectrical energy barriers, both ionic and nonionic surfactants are used and can serve as dispersing agents. A steric barrier to aggregation may be produced when the adsorbed surfactant molecules extend into the aqueous phase and inhibit the close approach of two particles to each other. Nonionic surfactants of the polyoxyethlene type are excellent dispersing agents for this purpose, because their highly hydrated polyoxyethylene chains extend into the aqueous phase in the form of coils that present steric barriers to aggregation [26].

Unsubstituted phthalocyanines are not solubilized or dispersed in the micellar solution of cationic surfactant with a ferrocenyl moiety. Phthalocyanines disperse in the micellar solution of nonionic surfactants with a ferrocenyl moiety. A nonionic surfactant with the ferrocenyl moiety used is α-(11-ferrocenyl-undecyl)-ω-hydroxy poly(oxyethylene) (12.3) (FPEG), which has been prepared by the following scheme [23, 25]:

$$FcH + BrC_{10}H_{20}COCl \xrightarrow{\ AlCl_3/CH_2Cl_2\ } FcCOC_{10}H_{20}Br$$

$$\xrightarrow{\ Zn/Hg\ } FcC_{11}H_{20}Br \xrightarrow{\ NaO(CH_2CH_2O)_nH\ } FPEG$$

Scheme I

where FcH is ferrocene. This surfactant is now commercially available [27]. FPEG is a yellow oil. Two main peaks of its UV-vis spectrum (ethanol) at 326 nm (ε=64) and 438 nm (100) are close to those of ferrocene at 325 nm (ε=50) and 440 nm (90) [25]. FPEG in an aqueous solution slowly decomposes under light in the presence of oxygen. Such a phenomenon has been reported in the case of methylferrocene in a dodecyl trimethylammonium bromideaqueous solution. However, FPEG in an aqueous solution is stable for a few months in the dark.

The values of critical micelle concentration (*CMC*) have been reported by the dye solubilization method (8 (0.2 M Li$_2$SO$_4$)[24, 25], 15 μM [28]) and the surface tension method (10 (0.2 M Li$_2$SO$_4$ [25]), 13 μM [28]). The difference in the value between two reports may be due to the difference in the salt effect [29]. These small values of the *CMC* are ascribable to the absence of any dissociable group [30] and the hydrophobic character of the ferrocenyl moiety. For example, ferrocene is fairly soluble in hexane and hardly soluble in water.

The micellar aggregation behavior of FPEG in water has been investigated using dynamic and static modes of light scattering, small-angle x-ray scattering, and transmission electron microscopic techniques by Yokoyama et al. [28]. The results of these measurements show that the micelle is a spherical particle with a radius of 3.8 ± 0.3 nm and with an aggregation number of 62 on the average.

C. ELECTROCHEMISTRY

The electrochemical behavior of ferrocene (FcH) and its derivatives has been reported by many researchers. Most of these studies have been examined in organic solvents owing to their poor solubilities in water [31]. They found that these compounds undergo a one-electron oxidation to form the unipositive ferrocenium ion (FcH+), and that the ferrocene /ferrocenium redox reaction is reversible:

$$FcH \underset{+e}{\overset{-e}{\rightleftharpoons}} FcH^+$$

In aqueous solutions, ferrocenes are also oxidized to ferrocenium ions [32]. Electrochemistry of ferrocene solubilized in micellar solution has been studied by Yeh and Kuwana [33] and Ohsawa and Aoyagui [34]. They show that ferrocene molecules are incorporated as micelles that readily transfer electrons with platinum electrode. Cyclic voltammetric data indicate that the electron-transfer reaction is also reversible.

Electrochemical and spectroscopic studies of cationic surfactants, (ferrcocenylmethyl)dodecyldimethylammonium bromide (I+) and (11-ferrocenyl)undecyltrimethyl-ammonium, having a ferrocenyl moiety, have been reported by us [35, 36]. The concentration dependence of the diffusion coefficient values for I+ and its oxidized form I^{2+} show that the I+ surfactants form redox-active micelles and that these micelles can be broken up into monomers by oxidation and reform by reduction.

The cyclic voltammogram (*CV*) of an aqueous solution containing 2 m*M* FPEG and 0.2 *M* (0.1 *M*) Li_2SO_4 (LiBr)

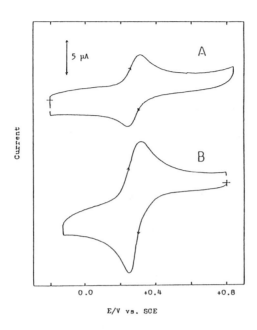

Figure 3 Cyclic voltammograms of aqueous solutions containing (A) 2.0 mM FPEG and 0.1 M LiBr, and (B) 2.0 mM FPEG$^+$ and 0.1 M LiBr, where FPEG$^+$ was prepared by bulk electrolysis at 25 $^\circ$C. Scan rate: 20 mV s^{-1}. Working electrode (glassy carbon) area: 0.02 cm^2. Reproduced with permission from [24].

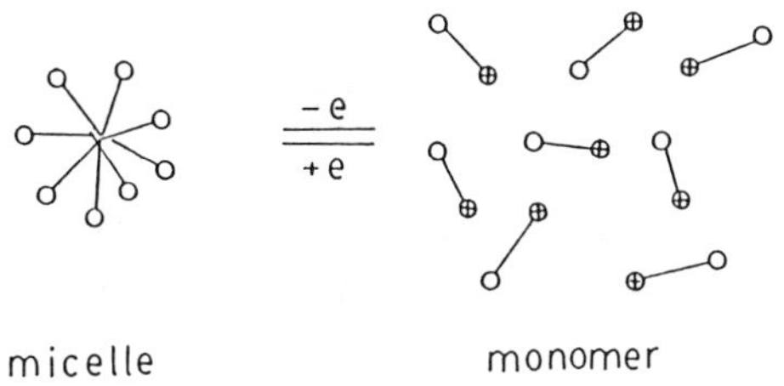

micelle   monomer

Figure 4 Schematic representation of reversible breakup and formation of micelles by redox reaction.

exhibits a quasireversible behavior with a half-wave potential ($E_{1/2}$) of +0.28 V (+0.21 V) versus SCE [Fig. 3(A)] [24]. This step has been assigned to the redox system FPEG/FPEG$^+$. Most FPEG may exist as micelles under these conditions since its critical micelle concentration is negligible (8 μM). The cyclic voltammogram of 2.0 mM FPEG$^+$ shows a reversible one-electron step with a $E_{1/2}$ of +0.28 V [Fig. 3(B)]. The solution of FPEG$^+$ was prepared by controlled potential bulk electrolysis of this solution.

When a step is reversible, the oxidation peak current (i_p) is given by the following relation [37-39]:

$$i_p = 269A\, n^{3/2}\, C_{red}\, D_{red}^{1/2}\, v^{1/2}$$

where A is the area of electrode, n the number of electrons, C_{red} (C_{ox} for reduction case) the concentration of reductant in the bulk solution, D_{red} (D_{ox} for the reduction case) diffusion coefficient of reductant, and v the scan rate.

The peak current for FPEG$^+$ is much larger than that for the FPEG solution. This finding suggests that the value of the diffusion coefficient of FPEG$^+$ is much greater than that of FPEG, due to the above relation. As the value of the diffusion coefficient is nearly proportional to the inverse of the square root of a molecular weight (M), $D \propto M^{-0.55}$ [40], the weight of FPEG is much lower than that of FPEG$^+$ in the aqueous solution. This suggests that most FPEG molecules exist as micelles, whereas all the FPEG$^+$ molecules exist as monomers (Fig. 4). These facts indicate that FPEG$^+$ may not have the ability to act as a surfactant to disperse pigments.

D. EXPERIMENTAL METHODS

i. Preparation of a Surfactant Solution Dispersing Pigment Particles

The dispersion of these pigments is ascribable to the following reasons [25]: First, the concentrations of these pigments are much larger than those in the case of solubilization in micelles. Second, the concentration depends on the size of the particles [for example, C_{disp} of

Table 1 Results of Film Formation Studies of Phthalocyanines Dispersed in 1 or 2 mM FPEG, 0.1 M LiBr Aqueous Solution

Compound	Size of particles (μm)	CFPEG (mM)	Cadded (mM)	Cdisp (mM)	Electrolysis time (min)	Q (mC cm^{-2})	Appearance[a]	Ref.
H2Pc (Tokyo Kasei)	0.1-0.2	2	10	3	30	16	greenish blue	[23, 53]
H2Pc (BASF)		2	20		30	8	greenish blue	c
H2PcCl16 (BASF)		2	15		30	12	green	c
H2(CN)8 (Toyo Inki)		2	10		70		green	c
H2Nc[b]		1	4		13 h		green	c
α-CuPc (Dainichiseika)	0.1-0.2	2	10	7	30	10	blue	[24]
β-CuPc(1) (Dainichiseika)		2	10		30	6	blue	[24]
β-CuPc(2) (Tokyo Kasei)	0.1-0.2	2	10	4	30	10	blue	c
β-CuPc(3) (Kanto Chemical Co.)	2-5	2	10	1	60		no film	[24]

ε-CuP (Toyo Inki)	0.05-0.1	2	10	4	20	7	blue	[24]
CuPcCl (Dainichiseika)	0.05-0.1	2	10	5	20	12	blue	[25]
CuPcCl16 (Tokyo Kasei)	0.05-0.1	2	20	11	20	6	bluish green	[25]
CuPcBr6Cl10 (BASF)	0.05-0.1	2	10	8	20	10	green	[25]
CuPcBr8Cl8 (BASF)	0.05-0.1	2	10	3	20	7	green	[25]
CuPc(CN)8 (Toyo Inki)		2	10		70		green	c
MgPc (Tokyo Kasei)	0.2-1	1	5		14 h		blue	[54]
AlPcCl (Nippon Shokubai)		1	5		180		blue	[54]
CoPc (Tokyo Kasei)		1	5		200		blue	c
NiPc (Kodak)		2	10		15 h		blue	c

a All of the films are transparent.
b Obtained from Dr. Shimura, Tokyo Metropolitan University.
c T. Saji, unpublished data.

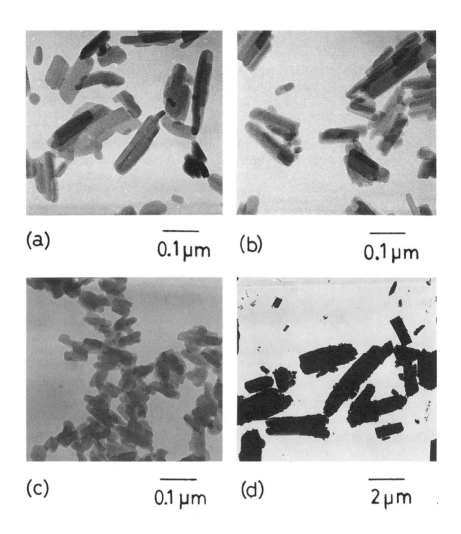

Figure 5 Transmission electron micrographs of pigment particles; (a) α-CuPc, (b) β-CuPc(1), (c) ε-CuPc, (d) β-CuPc(3).

β-CuPc(1) and (3) in Table 1]. Transmission electron microscopy images of some CuPc particles are given in Fig. 5. The sizes of the first three pigments [(a)-(c) in Fig. 5] are approximately 0.1-0.2 μm. These pigments disperse sufficiently. On the other hand, the size of β-CuPc (3) is much larger than that of the above pigments. This pigment disperses poorly. Generally, the smaller particles are advantageous for their dispersion due to their slow sedimentation. Third, the SEMs of the particles obtained by evaporation of these solutions show that the size and crystalline form of these particles agree with those of the particles added to the solution.

The behavior of the disperse system is described by taking into account the sedimentation and diffusion of the particles [41, 42]. The sedimentation velocity (*u*) is given by the Stokes equation:

$$u = 2a^2 (\rho - \rho_0)g/9,$$

where *a* is the diameter of the particle, $(\rho - \rho_0)g$ the difference in specific gravity, the coefficient of viscosity of the dispersing liquid. On the other hand, the mean displacement (*x*) during time (*t*) is given by

$$x = (RTt/3\pi a N_A)^{1/2},$$

where N_A is Avogadro number. The results of calculations of these equations show that the sedimentation is the predominating process when *a* >>1 µm, and Brownian movement is the predominating process when *a* << 0.1 µm [42]. Based on these results, the size of the particles used for the MD method should be smaller than 1 µm [42].

In order to disperse pigment particles using surfactant, an aqueous solution containing sufficient surfactant, supporting electrolyte, and pigment have to be sonicated and stirred. The periods of the sonication and the stirring required for film preparation depends on the properties of pigment [especially its surface area (size of particles) and aggregation force among pigment particles] and the concentrations of pigment and surfactant. Usually, the following method has been used for dispersing pigments and their film formations:

An aqueous solution containing 2.0 mM FPEG, 0.1 *M* LiBr, and 10 mM pigment was sonicated for 10 min and stirred for 3 d.

In our experimental results, a pigment film is not

formed in the following cases:

1. When the periods of the sonication and the stirring are not enough, a film is not formed due to the high concentration of surfactant unadsorbed on the particle surface (free surfactant) which disturbs the deposition of pigment particles.
2. When the size of pigment particles is too large (more than 1 μm), a film is not formed due to the same reason as in 1 and their sedimentation.
3. When the size of pigment particles is too small, the particles sediment soon after stirring the solution owing to an insufficiency of surfactant.
4. When cohesive forces between particles and the electrode are so weak, they are not held on the electrode. Sometimes, particles are washed with distilled water in order to remove unidentified impurities in commercial pigment, which influence these cohesive forces [25].

Table 1 lists the concentration of pigments (C_{dip}) that are dispersed by 2.0 mM FPEG in 0.1 M LiBr aqueous solution.

ii. Electrolysis Cell

Controlled-potential electrolysis has been carried out with a three-electrode cell. A simple electrolysis cell for film formation is illustrated in Fig. 6. A substrate (electrode) and a saturated calomel electrode (SCE) are used as a working and a reference electrode, respectively. As an auxiliary electrode, a large area of platinum or aluminum is used. The substrate is fixed by a metal clip, which makes the substitution of the substrate easier. Bubbling nitrogen through the solution is not required for a quick electrolysis (within 1 h). At the beginning of this experiment, the solution was stirred slowly; However, we can obtain a more uniform film without stirring the solution. For a large-area substrate (100 x 100 mm^2) a more simple electrolysis cell is used (Fig. 7). The salt bridge is inserted directly into the electrolytic solution.

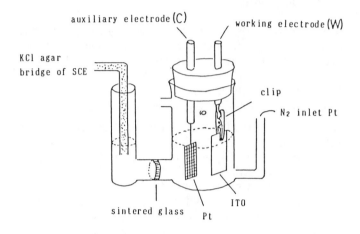

Figure 6 Electrolysis cell for film formation.

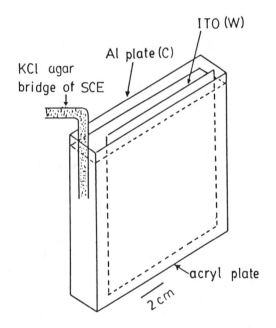

Figure 7 Simple electrolysis cell for formation of a large -area film.

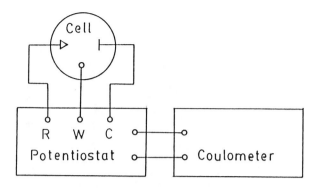

Figure 8 Electrochemical system for film formation: (R) Reference electrode, (W) working electrode, (C) auxiliary electrode.

E. FILM FORMATION

i. Phthalocyanines

The cyclic voltammogram of an aqueous solution containing 2.0 mM FPEG, 0.1 M LiBr, and 10 mM pigment [e.g., β-CuPc(1)] at the ITO electrode shows a quasireversible step, and the peak current is nearly 60% of that without pigment [25]. Such a decrease in peak current is ascribable to the formation of aggregates among FPEG and pigment particles. Our later experimental results indicate that most of the current is ascribable to free FPEG. As will be described in Section F, the pigment particles deposit when the concentration of the free FPEG decreases to less than the *CMC*. If this concentration is 1 mM, in order to oxidize 99% of FPEG by controlled-potential electrolysis, the electrode has to be maintained at a more positive potential by 120 mV than the standard potential (approximately half-wave potential) of FPEG/FPEG⁺ based on the Nernst equation:

$$E = E_0 + (RT/nF) \ln (C_{ox} / C_{red})$$

where E is the electrode potential, E_0 standard potential, and C_{ox} (C_{red}) the concentration of oxidant (reductant).

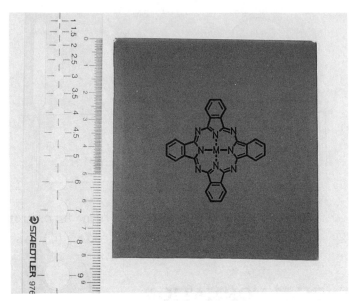

Figure 9 Photograph of large-area β-CuPc(1) film prepared by electrolysis of an aqueous solution containing 10 mM β-CuPc(I), 2.0 mM FPEG and 0.1 M LiBr at ITO for 20 min using the cell in Fig. 7.

Therefore, the potential at the controlled-potential electrolysis depends on the concentration of free FPEG. Usually, controlled-potential electrolysis for film formation has been done at +0.50 V versus *SCE* using a potentiostat with a coulometer for measurement of the amount of current through the electrode (Fig. 8).

In most research, indium tin oxide (ITO) has been used as an electrode (substrate). In addition to ITO, carbon, gold, and platinum are possible to use; Aluminum and copper are not used, due to their low oxidation potentials. The oxidation potential of the electrode must be higher than that of the electrolysis for film formation (approximately +0.5 V versus *SCE*).

A typical representative photograph of the thin films prepared by the MD method is shown in Fig. 9, which illustrates that this film has uniform thickness and transparency. The results of film formation are listed in Table 1. Most of the experiments have been done using an aqueous solution containing a 2 mM FPEG, 0.1 M LiBr, 10 mM pigment solution. Most of these pigment films have been

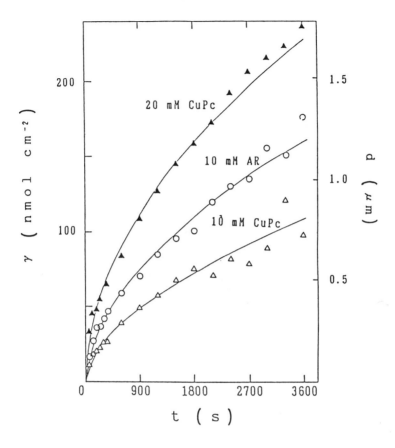

Figure 10 Film thickness (d) of β-CuPc(1) (CuPc) or dianthraqunoyl red (AR) versus electrolysis time. These films were prepared by electrolysis of an aqueous solution containing pigment, 2.0 mM FPEG, and 0.1 M LiBr at ITO [43].

formed by the MD method. No film is prepared for β-CuPc(3) [25]. This failure is ascribable to the low dispersibility of this pigment due to its large particle size (2-5 μm) and high concentration of free FPEG. In order to prepare a film of $CuPcCl_{16}$, the addition of a larger amount of the pigment is necessary [25].

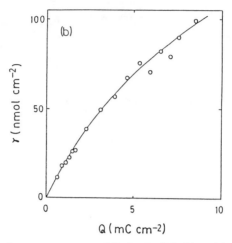

Figure 11 Plot of the amount of β-CuPc(1) film (γ) versus that of electricity passed through the electrode (Q). Reproduced with permission from [25].

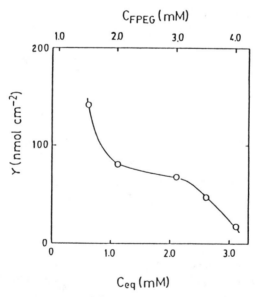

Figure 12 Amount of β-CuPc(1) film (γ) during 20 min electrolysis versus the concentrations of FPEG (C_{FPEG}) or equilibrium FPEG (C_{eq}) [45].

The film thickness change with electrolysis time for β-CuPc(1) and dianthrquinoyl red (AR) has been investigated by us [43]. The film thickness of β-CuPc increases with electrolysis time (Fig. 10). The rate of film growth is proportional to the amount of pigment added to the solution. A log plot analysis shows that it depends approximately on the square root of the electrolysis time, which suggests that the rate is controlled by the diffusion of pigment particles. Most of the films in Table 1 become more than 1 μm thick after 30-60 min electrolysis. The thickness of β-CuPc(1) increases to more than 10 μm during overnight electrolysis.

The amount of the β-CuPc(1) film deposited on a carbon electrode versus the electricity passed through the electrode has been reported (Fig. 11) [25]. This plot does not exhibit linearity, in contrast to that of azo dye solubilized in micelles. This is ascribable to the difference in the film forming process between these two cases. Most of the current passed through the electrode in the case of azo dye contributes to its film formation. On the other hand, the current in the case of pigment partially contributes to film formation, since free surfactant (monomer or micelle) does not participate in dispersing the pigment. The amount of the free surfactant changes with time.

The amount of the β-CuPc(1) film deposited during 20 min electrolysis versus the concentration of FPEG (or free FPEG) has been investigated (Fig. 12) [44]. The film is not formed at a concentration of more than 3.5 mM. This fact suggests that (1) the excess free surfactant disturbs the deposition of the pigment by electrolysis of the surfactant and (2) the lower concentration of the free surfactant gives better current efficiency.

ii. Other Pigments

Except for phthalocyanines, films of following pigments have been prepared by the MD method:

Dianthraquinoyl Red (Ciba-Geigy), Perylene Maroon (Bayer), Perylene Red (BASF), Perylene Scarlet (Bayer), Perylene Vermillion (BASF), Pyranthron Red (BASF) [25], 5,10,15,20-tetra-phenylporphyrin and its metal complexes [45], Cromophtal Blue (Ciba-

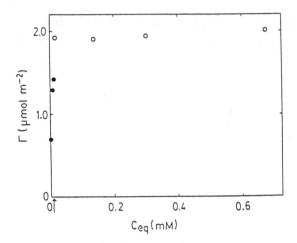

Figure 13 Adsorption isotherm of FPEG on β-CuPc(1) particles (10 mM): Closed circles indicate that these particles were sedimented without centrifugation. The arrow indicates the *CMC* of FPEG. Reproduced with permission from [25].

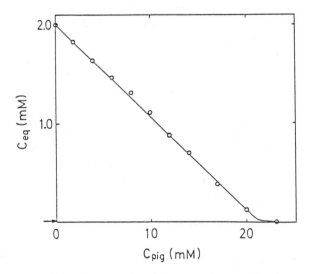

Figure 14 Plot of the equilibrium concentration of FPEG (C_{eq}) versus the amount of β-CuPc(1) particles (C_{pig}). Total concentration of FPEG was kept at 2.0 mM all throughout the experiment. Reproduced with permission from [25].

Geigy), Sumitone Green (Sumitomo), Fast Rose Lake B (Noma), Violet GI (Noma) [46].

F. MECHANISM OF FILM FORMATION

Adsorption Isotherm

A type of adsorption isotherm commonly observed in adsorption from solutions of surfactants is the Langmuir-type isotherm, expressed by the equation [47]:

$$\Gamma = (\Gamma_M K C_{eq})/(1 + K C_{eq})$$

where Γ is the surface concentration of free (equilibrium) surfactant, in moles m^{-2}, Γ_M the surface concentration of the maximum adsorbed surfactant at monolayer adsorption, C_{eq} the concentration of the free surfactant in the liquid phase at adsorption equilibrium, K an equilibrium constant of the surfactant between the surface and the liquid phase. The adsorption behavior of FPEG on β-CuPc(1) has been reported [25]. Figure 13 shows adsorption isotherm of FPEG on β-CuPc(1) particles, where the amount of the pigment are kept at 10 mM all through this experiment. The pigment particles in a low concentration of FPEG (closed circle in Fig.13) are sedimented without centrifugation. This sedimentation suggests that the pigment particles deposit when the concentration of free surfactant (C_{eq}) decreases to less than the *CMC*. The surface coverage is seen to increase with increasing concentration of the free surfactant until a saturation value of 2.0 μmol^{-2} (Γ_M) is attained at slightly above the critical micelle concentration of FPEG (8 μM). The value of Γ_M gives an area per molecule of 83 $\mathring{A}^2$ molecule^{-1}. This value is approximately close to those for similar surfactants [48-52], which provides evidence for monolayer adsorption at saturation.

The free concentration of FPEG (C_{eq}) versus the concentration of pigment (C_{pig}) added to a 2.0 mM FPEG aqueous solution has been reported [25]. The value of C_{eq}

decreases linearly with increasing amount of pigment (C_{pig}) (Fig. 14). The inclination of the straight line in this figure gives the same value of Γ_M as that obtained by the adsorption isotherm. The pigment particles are sedimented at 27 mM of C_{pig} , which indicates that their sedimentation occurs by a desorption of 20% of the surfactant of the surface coverage.

On the basis of these results, the mechanism of the film formation has been speculated as follows [25]:

1. The free surfactant (FPEG) diffuses to the electrode surface and is oxidized to its cation (FPEG$^+$). The concentration of the free surfactant in the vicinity of electrode decreases to less than the *CMC*.

2. The surfactants adsorbed on the pigment are desorbed from the pigment surface in order to satisfy the adsorption equilibrium. This desorption leads to the deposition of pigment on the electrode, which occurs efficiently when the concentration of the free surfactant decreases to less than approximately the *CMC*.

3. After the electrode is covered with the pigment film, the free surfactant diffuses in the film, owing to the existence of small space between the particles in the film, eventually reaches the electrode surface, and is electrolyzed. The concentration of the free surfactant in the vicinity of the film is kept at less than the CMC so that the film continues to grow for a long period to a reasonable thickness.

G. PROPERTIES OF FILMS

i. Morphology of Films

The scanning electron micrographs (SEM) of H$_2$Pc [23, 53], CuPc [24], and MgPc [54] prepared by this method have been reported. Figure 15 shows a typical SEM of α-CuPc, CuPcBr$_8$Cl$_8$, and H$_2$Pc. These photographs show these films to be composed of small particles. The size and crystal form of the particles in these films are the same as those of the added particles in the surfactant solutions, and do not

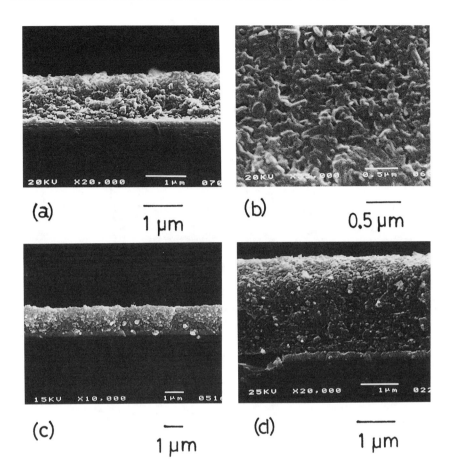

Figure 15 Scanning electron micrographs of cross sections and surfaces of the films prepared by the electrolysis of an aqueous solution containing pigment, 2.0 mM FPEG, and 0.1 M LiBr at the ITO electrode for (a), (b), (c) 30 min and (d) overnight, (a), (b) α-CuPc(1), (c) CuPcBr$_8$Cl$_8$, (d) H$_2$Pc.

change after 15 min, 1 h, or 10 h of electrolysis (e.g. H$_2$Pc [23], β-CuPc(1) [24]). The results of the x-ray-diffraction patterns of CuPc film and added particles have supported these facts [24]. The transparency of these films is ascribable to the fact that the size of these particles is smaller than 0.3 µm. Light scattering is negligible when the

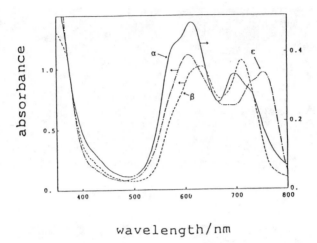

Figure 16 Electronic absorption spectra of CuPc film parepared by the electrolysis of an aqueous solution containing CuPc, 2.0 mM FPEG, and 0.1 M LiBr at the ITO electrode for 5 min, (α): α-CuPc, (β): β-CuPc, (ε): ε-CuPc. Reproduced with permission from [24].

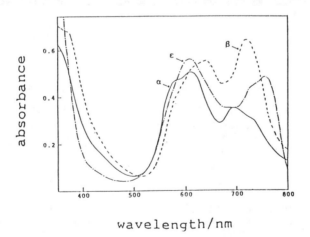

Figure 17 Electronic absorption spectra of aqueous solutions containing 20 mM CuPc, 2.0 mM FPEG, and 0.1 M LiBr; (α) α-CuPc, (β) β-CuPc, (ε) ε-CuPc. Reproduced with permission from [24].

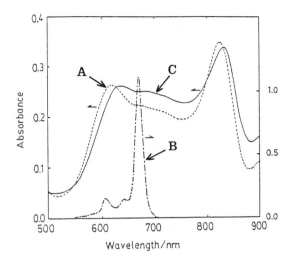

Figure 18 Electronic absorption spectra of MgPc in 1 m*M* FPEG, and 0.1 *M* LiBr aqueous solution (A), in ethanol solution (B), and MgPc film prepared by electrolysis of an aqueous solution containing 5 m*M* MgPc and 0.1 *M* LiBr at the ITO electrode maintained at +0.5 V versus SCE for 1 h (C). Reproduced with permission from [54].

size of particle is less than half the wavelength of visible light. The SEM of the cross sections of the films in Fig. 15 show that these are of uniform thickness and that the film thicknesses are approximately 1 μm.

ii. Spectral Properties of Film

Color photographs of films of α-, ε-CuPc, $CuPcBr_6Cl_{10}$ have been reported [55]. The films prepared by this method have original pigment colors and are transparent. The absorption spectra of α-, β-, ε-CuPc prepared by this method have been reported [24]. Figure 16 shows the absorption spectra of CuPc films on the ITO electrode. These spectra consist of broad peaks and are very similar to those for the corresponding surfactant solution (Fig. 17).

The spectra of the films prepared from α- and β-CuPc particles are very similar to those of α- and β-CuPc films prepared by vacuum sublimation [11, 12, 14]. The spectrum of the film prepared from ε-CuPc particles is also similar to that of the ε-CuPc polyester film [56]. The agreement of absorption spectra and x-ray-diffraction patterns among CuPc particles, their micellar solution, and their films indicate that the crystalline form of the pigment is maintained throughout the film preparation processes.

The spectral properties of the films of MgPc and AlClPc have been reported (Fig. 18) [53]. They show an intense absorption peak in the near-infrared region and have been proposed as candidates for laser recording materials [13, 56-58]. The formation of such films without treating the films with organic solvents is relevant to the process of preparing these films as previously described .

iii. Electrochemical Properties

Photoelectrochemical characteristics of a H_2Pc electrode prepared by the MD method have been reported by Harima et al. [59]. This technique has been found to provide more photoactive phthalocyanine layers on the ITO, compared with a vacuum sublimation technique. A short-circuit photocurrent of 0.1 mA cm^{-2} was obtained for the ITO/H_2Pc/I_3^{-1},I^{-1}/Pt cell under the white light illumination of 6 mW cm^{-2}, together with an open-circuit photovoltage of 70 mV and a fill factor of 0.42. These data lead to a value of 0.06% for the energy conversion efficiency.

The electrochemical characterization of H_2Pc thin film prepared by the MD method has been reported by Harima et al. [52]. This film on the ITO electrode has been investigated mainly by use of an electrochemical technique. It has been found that the Pc layer has a high electrolyte permeability. This fact has been ascribed to the structure of the film, which consists of particles folded loosely with each other.

H. CONCLUSIONS

In conclusion, we have discussed the experimental practice, the principle, and the results of the electrochemical formation of phthalocyanine thin films by the MD method. The advantages of this technique for the film formation of pigments such as phthalocyanines are as follows:

1. This technique enables thin films of a wide variety of pigments to be prepared that satisfy following conditions: (a) the pigments are dispersible in a surfactant solution and (b) they are not electrolyzed at the potential for oxidation of the surfactant with ferrocenyl moiety (+0.5 V versis *SCE*).

2. The starting pigments do not undergo electrochemical reactions when their films are formed; hence film-forming pigments are the same as the starting pigments. Generally, electrochemical film formation proceeds via electrochemical reactions of organic compounds and are different from those of the starting compounds.

3. The crystalline form of the films is easily controlled. The crystalline form of the starting pigment is maintained throughout the film preparation processes.

4. Film thickness is easily controlled by the amount of the pigment added to the solution, and the period of electrolysis time.

5. This technique enables a large surface to be coated with a uniform thin film.

6. A large number of films are prepared by one pigment solution. This is an advantage for industrial production.

The MD method has the same limitations as electrochemical film formation in that (1) the substrate has to be an electrical conductor, and (2) the oxidation potential of the substrate has to be more positive than +0.5 V versus *SCE*. These limitations may be overcome by an electroless method. Since the elapsed time since the invention of the MD method is relatively brief, the thin films prepared by this technique have not been well characterized. Therefore, much experimental work remains to be done.

ACKNOWLEDGMENTS

We thank R. Ohoki for the electron micrograph data. This work partially supported by a Grant-in-Aid for Scientific Research from the Ministry of Education, Science and Culture (Nos. 01604540 and 02205046), the Kurata Foundation, and the Matsuda Foundation.

REFERENCES

1. A. B. P. Lever, *Adv. Inorg. Radiochem.*, 7 (1965) 27.
2. L. J. Boucher, *Coordination Chemistry of Macrocyclic Compounds*, G. A. Melson, Plenum Press, New York, Chap. 7.
3. K. Kasuga and M. Tsutsui, *Coord. Chem. Rev.*, 32 (1980) 67.
4. Ref. [1], p. 92.
5. A. Giraudeau, F. F. Fan, and A. J. Bard, *J. Am. Chem. Soc.*, 102 (1980) 5137.
6. V. R. Shepard, Jr. and N. R. Armstrong, *J. Phys. Chem.*, 83 (1979) 1268.
7. J. M. Green and L. R. Faulkner, *J. Am. Chem. Soc.*, 105 (1983) 2950.
8. A. W. Snow and W. R. Barger, *Phtalocyanines*, C. C. Leznoff and A. B. P. Lever, Eds., VCH, New York, 1989, Chap. 5.
9. C. W. Tang, *Appl. Phys. Lett.*, 48 (1986) 83.
10. G. Perrier and L. H. Dao, *J. Electrochem. Soc.*, 134 (1987) 1148.
11. E. A. Lucia and F. D. Verderame, *J. Chem. Phys.*, 48 (1968) 2674.
12. J. H. Sharp and M. Abkowitz, *J. Phys. Chem.*, 77 (1973) 477.
13. C. H. Griffiths, M. S. Walker, and P. Goldstein, *Mol. Liq. Cryst.*, 33 (1976) 149.

14. P. S. Vincett, Z. D. Popovic, and L. McIntyre, *Thin Solid Films*, 82 (1981) 357.

15. A. Taomoto, Y. Machida, K. Nichogi, and S. Asakawa, *Nippon Kagakukaishi* (1987) 2025.

16. A. E. Alexander, *J. Chem. Soc.* (1937) 1813.

17. Ref. [8], pp. 367-78.

18. K. Hoshino and T. Saji, *J. Am. Chem. Soc.*, 109 (1987) 5881.

19. K. Hoshino and T. Saji, *Chem. Lett.* (1987) 1439.

20. K. Hoshino, M. Goto, and T. Saji, *Chem. Lett.* (1987) 547.

21. K. Hoshino and T. Saji, *Nippon Kagakukaishi* (1990) 1014.

22. T. Saji, *Yukagaku*, 39 (1990) 717.

23. T. Saji, *Chem. Lett.* (1988) 693.

24. T. Saji and Y. Ishii, *J. Electrochem. Soc.*, 136 (1989) 2953.

25. T. Saji, K. Hoshino, Y. Ishii, and M. Goto, *J. Am. Chem. Soc.*, 113 (1991) 450.

26. M. J. Rosen, *Surfactants and Interfacial Phenomena*, Wiley, New York, 1978, pp. 266-69.

27. Dojindo Laboratories, Kumamoto, Japan.

28. S. Yokoyama, H. Kurata, Y. Harima, K. Yamashita, K. Hoshino, and H. Kokado, *Chem. Lett.* (1990) 343.

29. K. Shinoda, T. Nakagawa, B. Tamamushi, and T. B. Isemura, *Colloidal Surfactants*, Academic Press, New York, 1963, pp. 139-41.

30. Ref. [29], pp. 133-41.

31. C. K. Mann and K. K. Barnes, *Electrochemical Reactions in Nonaqueous Systems*, Marcel Dekker, New York, 1970, Chap. 13.

32. H. M. Koepp, H. Wendt and H. Strehlow, *Z. Elecktrochem.*, 64 (1960) 483.

33. P. Yeh and T. Kuwana, *J. Electrochem. Soc.*, 123 (1976) 1334.

34. Y. Ohsawa and S. Aoyagui, *J. Electroanal. Chem.*, 114 (1980) 235.

35. T. Saji, K. Hoshino, and S. Aoyagui, *J. Am. Chem. Soc.*, 107 (1985) 6865.

36. T. Saji, K. Hoshino, and S. Aoyagui, *J. Chem. Soc., Chem. Commun.*, (1985) 865.

37. H. Matsuda and Y. Ayabe, *Z. Elektrochem.*, 59 (1955) 494.

38. R. S. Nicholson and I. Shain, Anal. Chem., 36 (1964) 706.

39. A. J. Bard and L. R. Faulkner, *Electrochemical Methods*, Wiley, New York, 1980, p. 218.

40. C. Tanford, *The Physical Chemistry of Macromolecules*, Wiley, New York, 1961, p. 312.

41. A. Sheludko, *Colloid Chemistry*, Elsevier, Amsterdam, 1966, pp. 55-61.

42. A. Kitahara and K. Furusawa, *Bunsan Nyukakei no Kagaku*, Kougaku Tosho, Tokyo, 1979, pp. 3-7.

43. M. Goto, T. Sugimoto, and T. Saji, unpublished data.

44. F. Takeo and T. Saji, unpublished data.

45. T. Saji, unpublished data.

46. K. Takeda, Y. Harima, S. Yokoyama, and K. Yamashita, *Jap. J. Appl. Phys.*, 28 (1989) 204.

47. Ref. [26], pp. 36-38.

48. B. Kronberg, L. Kall, and P. Stenius, *J. Disp. Sci. Tech.*, 2 (1981) 215.

49. D. N. Furlong and J.R. Aston, *Colloids and Surfaces*, 4 (1982) 121.

50. B. Kronberg, P. Stenius, and Y. P. Thorssell, *Colloids and Surfaces*, 12 (1984) 113.

51. S. Partyka, S. Zaini, M. Lindheimer, and B. Brun, *Colloids and Surfaces*, 12 (1984) 255.

52. Th. V. D. Boomgaard, Th. F. Tadros, and J. Lyklema, *J. Colloid Interface Sci.*, 116 (1987) 8.

53. Y. Harima and K. Yamashita, *J. Phys. Chem.*, 93 (1989) 4184.

54. T. Saji, *Bull. Chem. Soc. Jpn.*, 62 (1989) 2992.

55. T. Saji and M. Goto, *Kagaku (Chemistry)*, 45 [10] (1990) color pages.

56. A. M. Hor and R. O. Loutfy, *Thin Solid Films*, 106 (1983) 291.

57. K. Arishima, H. Hiratsuka, A. Tate, and T. Okada, *Appl. Phys. Lett.*, 40 (1982) 279.

58. L.H. Dao and G. Perrier, *Chem. Lett.*, 1259 (1986) 1259.

59. Y. Harima, K. Yamashita, and T. Saji, *Appl. Phys. Lett.*, 52 (1988) 1542.

5

Catalytic Functions

and Application

of Metallo-

phthalocyanine

Polymers

Kenji Hanabusa and
Hirofusa Shirai

CATALYTIC FUNCTIONS AND APPLICATION OF METALLOPHTHALOCYANINE POLYMERS

Kenji Hanabusa and Hirofusa Shirai

Metallophthalocyanines (Mpc) are very stable metal complexes of macrocyclic tetraazaporphyrin. A metal ion of Mpc is present in a specific conjugated π electron environment similar to hemes *in vivo*, and Mpc therefore has many enzymelike unique functions. However, the poor solubility of Mpc results in major problems in utilizing these functions. Improvements in the solubility of Mpc are achieved by the introduction of functional groups and attachment to polymers. In this chapter, the enzymelike catalytic functions of Mpc-containing polymers as well as Mpc derivatives and their applications are described.

A. CATALYTIC FUNCTIONS AND APPLICATIONS

Metallophthalocyanine derivatives are characterized by various properties:

The aromatic π electrons resonate through the tetraazaporphyrin ring, in the center of which various metal ions can be coordinated;
The structure is highly conjugated planar and has two axial coordination sites for catalytic reaction;
The aromatic rings act as both donor and acceptor; and
It is very stable thermally.

The catalysis of Mpc for organic reactions is summarized in Table 1. Actually, the catalysis of Mpc proceeds enzymatically, as can be expected from a similar structure to metalloporphyrin in heme-enzyme and coenzyme vitamin B12. Therefore, the study on the catalysis of Mpc is of interest in connection with enzyme reactions, particularly with regard to the applications of enzymelike catalytic functions.

Table 1 Catalytic Reactions by Metallophthalocyanines

Type of reaction	Instance	Central metal ion
Oxidation	$C_6H_5-CHO + O_2 \longrightarrow C_6H_5-COOOH \xrightarrow{\quad C_6H_5-CHO \quad} 2\,C_6H_5-COOH$	Fe(III), Co(II)
	$CH_3-\overset{O}{\underset{O}{\diamond}} \xrightarrow{O_2} CH_3COOCH_2CH_2OH$	Fe(III), Cu(II)
	$C_6H_5-CH_2CH_3 \longrightarrow C_6H_5-CH=CH_2 + H_2$	Cr(III), Ni(II), Co(II)
	$H_2S + 1/2\,O_2 \longrightarrow S + H_2O$	Fe(III), Co(II), Cu(II)
	$2\,RSH + 2\,OH^- \longrightarrow 2\,RS^- + 2\,H_2O$	Co(II), Fe(III)
	$2\,RS^- + 2\,H_2O + O_2 \longrightarrow RSSR + H_2O_2 + 2\,OH^-$	Co(II), Fe(III)
	$\overset{CH(CH_3)_2}{\bigcirc} + O_2 \longrightarrow \overset{C(OOH)(CH_3)_2}{\bigcirc} \longrightarrow \overset{OH}{\bigcirc} + CH_3COCH_3$	Co(II)
	$\overset{-OH}{\underset{OCH_3}{\bigcirc}} \xrightarrow{H_2O_2} \text{products}$	Fe(III)
	$CH_3(CH_2)_4CH_3 \longrightarrow CH_2=CH(CH_2)_3CH_3 + H_2$	Pt(II), Cr(III)
	$CH_3\underset{OH}{CH}CH_2CH_3 \longrightarrow CH_3COCH_2CH_3 + H_2$	Zn(II), Pt(II), Ni(II)

Table 1 (*continued*)

type of reaction	instance	central metal ion
Reduction	$CH\equiv CH \longrightarrow CH_2=CH_2$	Co(II)
	quinoline $+ 2 H_2 \longrightarrow$ 1,2,3,4-tetrahydroquinoline	Sn(IV)
	pyrrole $+ 4 H_2 \longrightarrow CH_3CH_2CH_2CH_3 + NH_3$	Co(II)
	$C_6H_5-CH(Cl)-C(=O)-C_6H_5 + H_2 \longrightarrow C_6H_5-CH_2-C(=O)-C_6H_5 + HCl$	Fe(III)
	$CH_3SH + H_2 \longrightarrow 1/2 C_2H_6 + H_2S$	Pt(II), Pd(II), Cr(III)
Carbonylation	$2\ C_6H_5-Br + CO \longrightarrow C_6H_5-CO-C_6H_5 + Br_2$	Pd(II)
Decomposition	$2 H_2O_2 \longrightarrow 2 H_2O + O_2$	Fe(III), Co(II)
	$NH_2NH_2 \longrightarrow N_2 + 2 H_2$	Co(II), Cu(II), Ni(II) Mn(III), Fe(III)
Dehalogenation	$C_6H_5-CH(Cl)-C(=O)-C_6H_5$ + (N-benzyl-1,4-dihydronicotinamide) $\longrightarrow C_6H_5-C(=O)-CH_2C_6H_5$ + (N-benzylnicotinamide) Cl^-	Fe(III), Cu(II)
Decarbonylation	furfural (furan-CHO) $\longrightarrow$ furan $+ CO$	Fe(III), Co(II), Cu(II)
Polymerization	$n\ CH_2=C(CH_3)(COOCH_3) \xrightarrow{O_2} -(CH_2-C(CH_3)(COOCH_3))_n-$	Fe(III)
	n (2,6-dimethylphenol) $+ n/4\ O_2 \longrightarrow$ poly(2,6-dimethylphenylene oxide)$_n$ $+ n/2\ H_2O$	Cu(II), Fe(III)
Friedel-Crafts reaction	$C_6H_6 + RCl \longrightarrow C_6H_5-R + HCl$	Fe(III)

B. OXYGEN CARRIER

Homoglobin and myoglobin are heme proteins with protoheme IX **1** as the active site. They act as carriers and storage media for molecular O_2 *in vivo*. By mixing metallophthalocyanine tetrasulfonic acid **2** with apomyoglobin, which was prepared by removal of **1** from myoglobin, **2**-containing apomyoglobin was obtained [1]. The visible spectrum of this artificial enzyme was characterized by the disappearance of the Soret band at 418 nm, which was attributed to free heme protein, and by the appearance of Q bands of **2** bound to apomyoglobin, which were at 650 nm for **2**-Fe(III), 680 nm for **2**-Co(II), and 670 and 700 nm for **2**-Cu(II). The circular dichroism (CD) spectra indicated that α-helix contents increased up to 18% for **2**-Fe(III) and 22% for **2**-Co(II) by insertion of **2** to apomyoglobin. However, satisfactory results as an oxygen carrier were not obtained, because of instability of **2** bound to apomyoglobin.

1

M = Fe(III), Co(II), Cu(II)

2

Co(II)-phthalocyanine-containing polymer **3** was synthesized by the Friedel-Crafts reaction of Co(II)-octachloroformylphthalocyanine with a random copolymer of 2-vinylpyridine and styrene [2]. In aqueous solution, Co(II)-phthalocyanine in **3** is of a five-coordinate high-spin type, and one axial coordination site is occupied by an N atom of a pyridyl group. Such a geometry would seem favorable for the formation of an O_2 complex. Actually, superoxide ion O_2^-, which was generated by the reduction of an O_2 complex accompanying the oxidation of Co(II) ion, was detected in O_2-saturated aqueous solutions by ESR spectroscopy. These results indicate the potential of Co(II)-phthalocyanine as an O_2 carrier.

m = 0.50, p = 0.43, q = 0.07

3

C. OXIDASELIKE OXIDATION

Oxidation *in vivo* is generally performed by a series of reactions of oxidation-reduction enzymes, and some oxidation reactions are shown in Eqs. (1) and (2), which indicate the dehydrogenation of substrate (SH_2) by hydrogen-acceptors like NADH and O_2, respectively. The production of O_2^- by **3**, as mentioned in Section B, gives the impression that metallophthalocyanine derivatives catalyze the reaction of Eq. (2).

$$SH_2 + A \xrightarrow{\text{dehydrogenase}} S + AH_2 \tag{1}$$

$$SH_2 + O_2 \xrightarrow{\text{oxidase}} S + H_2O_2 \tag{2}$$

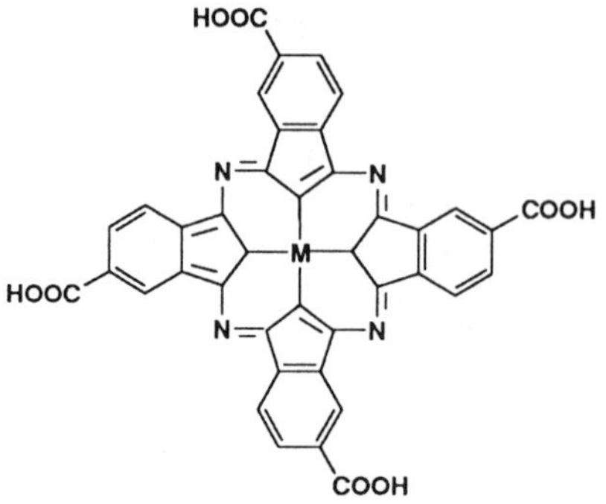

4 M = Fe(III)

5 M = Co(II)

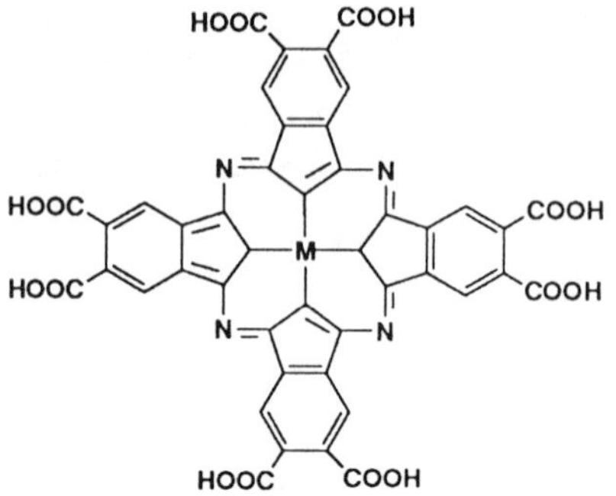

6 M = Fe(III)

7 M = Co(II)

Oxidaselike oxidation of 2-mercaptoethanol (RSH), shown in Eq. (3), was studied using Fe(III), Co(II)-phthalocyanine tetracarboxylic acids **4,5** and octacarboxylic acids **6,7** as catalyst [3]. The observed reaction rate at 25°C with *p*H 7 obeyed Eq. (4), and the proposed reaction mechanism is shown in Figure 1.

$$2HOCH_2CH_2SH + O_2 \xrightarrow{\text{catalyst}}$$

$$HOCH_2CH_2S\text{-}SCH_2CH_2OH + H_2O + 1/2O_2 \quad (3)$$

$$\frac{d[RSSR]}{dt} = k[Mpc]^{1.0}[RSH]^{2.0}[O_2]^{0.5} \quad (4)$$

First of all, RS$^-$, which was formed by dissociation of RSH in aqueous solution, coordinates to metallophthalocyanine from one axial coordination side, and then O_2 coordinates from the opposite coordination site. Consequently, ternary complex consisting of metallophthalocyanine, RS$^-$, and O_2 is formed in cycle A of Fig. 1, where K_{RSH}^H is the dissociation constant of RSH, k_1 and k_2 are the rate constants of coordination of RS$^-$ to Mpc and that of the reverse reaction, and k_3 and k_4 are the rate constants of coordination of O_2 as the sixth axial ligand and that of the reverse reaction.

Second, an electron transfer from RS$^-$ to O_2 *via* the central metal ion of the ternary complex occurs and results in the formation of thiol radical RS$\cdot$ and

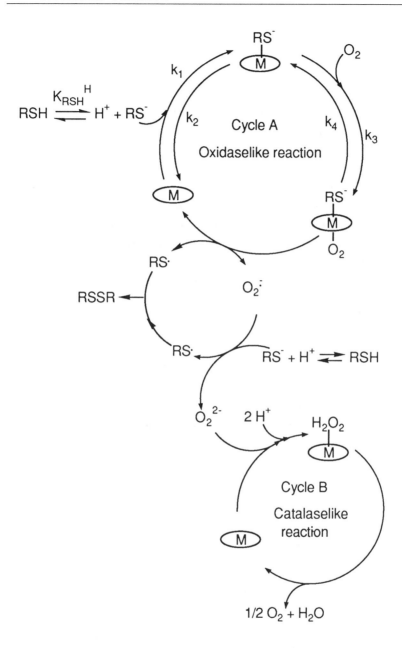

Figure 1 Mechanism of oxygen oxidation of thiol.
RSH: thiol; M: Mn(III), Fe(III), Co(II).

superoxide ion $O_2^{\cdot}$. This electron transfer expressed as a rate constant k_5 is the rate-determining step.

Finally, the reaction of RS^- and $O_2^{\cdot}$ produces $RS^{\cdot}$ and O_2^{2-}. The former affords to the final product RSSR by the coupling with $RS^{\cdot}$, while the latter decomposes to O_2 and H_2O by a catalaselike reaction in cycle B of Fig. 1. These steps are fast as compared with the electron transfer in cycle A. The use of the stationary state method about two equilibrium reactions (rate constants; k_1, k_2, k_3, and k_4) gives Eqs. (5) and (6), where V_0 is the initial velocity (mol 1^{-1} min^{-1}), V_{max} is the maximum velocity (mol 1^{-1} min^{-1}), which is equal to $k_5[Mpc]_{total}$, $K_A = (k_2k_5 + k_2k_4)/k_1k_5$, $K_B' = K_B[H^+]/K_{RSH}^H$, and $K_B=k_5/k_1$. Actually, good linear relationships between $1/V_0$ and $1/[O_2]$, or $1/V_0$ and $1/[RSH]$, were obtained. Kinetic data for the oxidation of 2-mercaptoethanol are summarized in Table 2. The activities of **7**, **6**, **5**, and **4**, which decrease in that order, indicate that metallophthalocyanine octacarboxylic acids are more active than metallo-phthalocyanine tetracarboxylic acids and Co(II) complexes are more effective than the corresponding Fe(III) ones.

Table 2 Kinetic Data for Oxidation of 2-Mercaptoethanol by Catalysts 3-7[a]

Catalyst	V_{max} (mol 1^{-1} min^{-1})	K_B (mol 1^{-1})	k_5 (min^{-1})	$\Delta H^{\neq b}$ (kcal mol^{-1})	$\Delta S^{\neq}$ (cal mol^{-1} deg^{-1})
3	2.21 x 10^{-4}	4.48 x 10^{-4}	2210	+5.69	-32.2
4	6.50 x 10^{-5}	5.90 x 10^{-3}	648	+9.46	-22.0
5	6.15 x 10^{-4}	4.74 x 10^{-3}	615	+8.26	-26.2
6	7.20 x 10^{-5}	2.47 x 10^{-3}	720	+7.10	-29.7
7	8.40 x 10^{-3}	5.05 x 10^{-3}	840	+4.59	-37.9

[a] $[M\text{-}pc]_0 = 2$ x 10^{-10} mol 1^{-1}, $[RSH]_0 = 7.13$ x 10^{-3}- 2.85 x 10^{-1} mol 1^{-1}, pH = 7.0.
[b] 10 - 40°C.

The kinetic data of Co(II)-phthalocyanine-containing polymer **3** is shown in Table 2. The value of k_5 for the heterogeneous oxidation of 2-mercaptoethanol by **3** was 2210 min^{-1}, which is about 3 times those of low molecular catalysts 4-

7. It is thought that the bonding on polymeric support causes the dissociation of the dimeric Co(II)-phthalocyanine segment into a monomeric one, which is essential for the coordination of RS⁻ and O_2, and also the subsequent electron transfer.

$$\frac{1}{V_0} = \frac{1}{V_{max}} + \frac{K_A}{V_{max}} \frac{1}{[O_2]} \tag{5}$$

$$= \frac{1}{V_{max}} + \frac{K_B'}{V_{max}} \frac{1}{[RSH]} \tag{6}$$

D. CATALASELIKE REACTION

Catalase consisting of apoprotein (MW 220,000-250,000) and four molecules of protoheme IX is widely distributed in nature. It catalyzes the decomposition of toxic H_2O_2, which is formed by oxidase, into O_2 and H_2O as in Eq. (7).

$$2H_2O_2 \xrightarrow{\text{catalase}} 2H_2O + O_2 \tag{7}$$

The reaction rate is very fast and is estimated to be about 6×10^6 molecule sec^{-1}. It is also known that the active center of catalase is a high-spin Fe(III) complex.

With these facts in mind, Fe(III)-phthalocyanine derivatives **4** and **6** were tested as a catalyst for the decomposition of H_2O_2 [4]. In aqueous solution at $pH>7$, **4** and **6** decomposed H_2O_2 *via* catalaselike reaction, as described in Fig. 2. By using the stationary state method of Michaelis-Menten, followed by Eadie-Hofstee method, were obtained Eqs. (8) and (9), where $K_m = (k_2 + k_3)/k_1$ and $V_{max} = k_3[\text{Fe(III)pc}]_{total}$. The electron transfer process (k_3) is the rate-determining step, and it can be determined by Eq. (9). The observed values of k_3 for **1**, **4**, and **6** in homogeneous aqueous solution at pH 7 were 6, 19, and 160 min^{-1}, respectively. The unexpectedly high value of k_3 for **6**, which is about 26 times that of natural material **1**, may be attributed to the fact that eight carboxy groups of **6** diminish the formation of dimer and oligomer by electrostatic and steric repulsion.

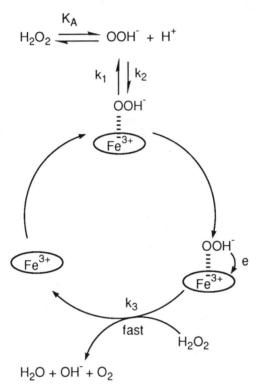

Figure 2 Mechanism of catalaselike reaction by Fe(III)pc.

$$V_0 = - \frac{V_{max}[H_2O_2]}{K_m + [H_2O_2]} \tag{8}$$

$$\frac{V_0}{[H_2O_2]} = - \frac{1}{K_m} V_0 + \frac{1}{K_m} V_{max} \tag{9}$$

Fe(III)-phthalocyanine-containing polymer **8** was prepared by the Friedel-Crafts reaction of Fe(III)-tetrachloroformylphthalocyanine with poly(styrene) instead of apoprotein, and its catalaselike activity was studied [5, 6]. The maximum velocity V_{max} of **8** rose to about 2 times that of **4**, and the activation energy fell to about half. In general, metallophthalocyanine ring tends to form a dimer, as shown Eqs. (10) and (11). However, the aggregation of metallophthalocyanine bonding to poly(styrene) like **8** becomes difficult due to the steric hindrance of both neighboring phenyl group and polymer chain. This may be the reason why **8** is more favorable as a catalyst than **4**.

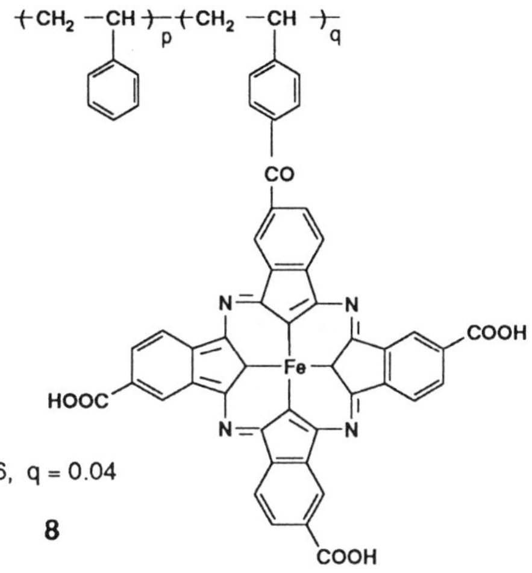

p = 0.96, q = 0.04

8

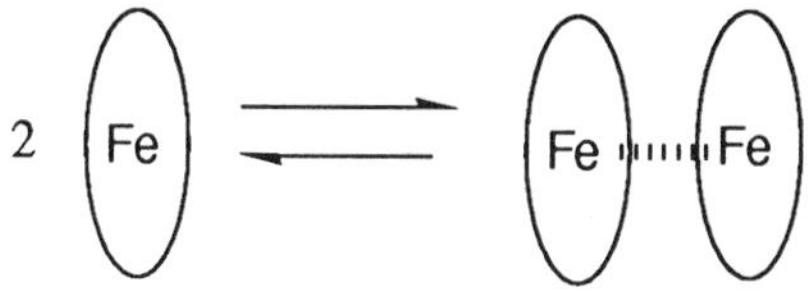

$$2 \; \text{Fe} \;\; \rightleftharpoons \;\; \text{Fe} \cdots \text{Fe} \qquad\qquad (10)$$

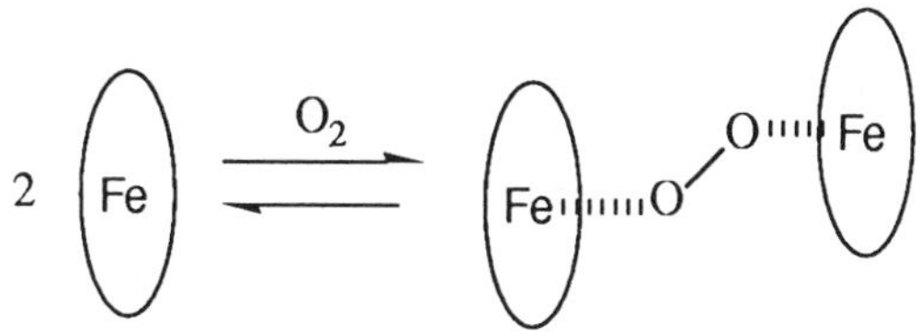

$$2 \; \text{Fe} \;\; \underset{}{\overset{O_2}{\rightleftharpoons}} \;\; \text{Fe} \cdots O \diagdown O \cdots \text{Fe} \qquad\qquad (11)$$

Protoheme IX **1**, that is, the active center in catalase, is surrounded by basic amino acid residues like lysine, and its axial coordination site is occupied by an N atom of an imidazole from a histidine residue. When the opposite axial site of protoheme IX is occupied by OOH⁻, which is generated by dissociation of H_2O_2, it decomposes through the exchange of an electron with another H_2O_2. It is believed that basic amino acid residues accelerate the dissociation of H_2O_2 and imidazolyl group of histidine builds up the five-coordinate high-spin Fe(III) complex suitable for the coordination of OOH⁻.

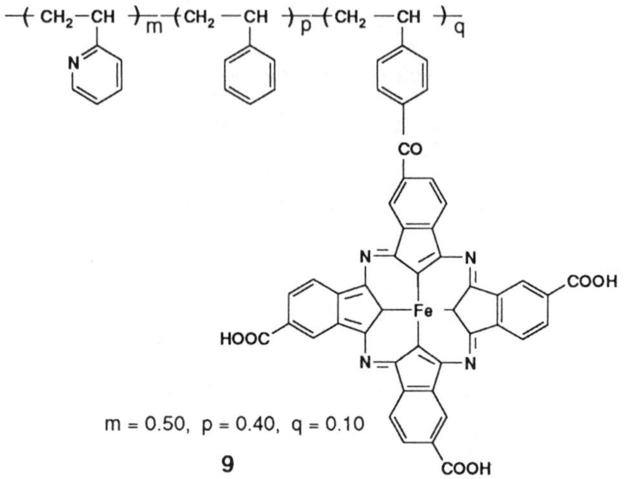

m = 0.50, p = 0.40, q = 0.10

9

Oxidaselike catalysis was examined by Fe(III)-phthalocyanine-containing copolymer **9** [2]. Polymer **9** has Fe(III)-phthalocyanine and a pyridyl group instead of protoheme IX and an imidazole group in catalase. The ESR spectroscopy of **9** supported a five-coordinate high-spin Fe(III)-phthalocyanine and the visible spectrum agreed with that of the monomeric Fe(III) one. The kinetic data for catalaselike H_2O_2 decomposition by polymer **9** are comparable to those of free Fe(III)-phthalocyanine compounds and catalase shown in Table 3. The value of V_{max} for **9** is 20 times greater than that for **4**, and the turnover number k_3 is about 10 times greater. Therefore, it can be said that Fe(III)-phthalocyanine bonding on the copolymer of styrene and 2-vinylpyridine is a remarkably effective catalyst. When rayon fiber having 14% crystallinity was treated with an alkaline solution of **6** and then acidified with HCl, a dark green fiber dyed with **6** was obtained [7]. Measurements on this compound **6** adsorbed on rayon fiber with a x-ray microanalyzer and microscopic electron spectroscopy indicate that monomeric **6** was dispersed homogeneously and deep into the interior of the fiber. The ESR spectrum of **6** adsorbed on rayon fiber is shown in Fig. 3. The g value of 4.32 can be attributed to the high-spin Fe(III) complex, which is an octahedral structure distorted along the Z axis. The kinetic data for the catalaselike reaction by **6** adsorbed on rayon is also given in Table 3. It is noteworthy that its k_3 is greater than that of **6**.

Table 3 Kinetic Data for Catalaselike Reaction by Phthalocyanine Analogues

Catalyst	V_{max} (mol l^{-1} min^{-1})	K_m (mol l^{-1})	k_3 (min^{-1})	E_a (kJ mol^{-1})
4[a]	5.2×10^{-4}	6.6×10^{-3}	22	17.6
6[a]	8.58×10^{-3}	7.05×10^{-3}	160	
8	1.0×10^{-3}	2.8×10^{-2}	60	7.5
9[b]	1.2×10^{-2}	2.2×10^{-2}	200	
6 adsorbed on rayon fiber	8.93×10^{-3}	7.05×10^{-3}	179	
Hemin			35	
Catalase			5×10^6	

[a] Homogeneous system, $[Fepc]_0 = 5 \times 10^{-5}$ mol l^{-1}, $pH = 7.0$.

[b] Heterogeneous system, $[Fepc]_0 = 5 \times 10^{-5}$ mol l^{-1}, $pH = 7.0$.

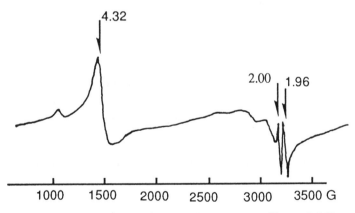

Figure 3 ESR spectrum of **6** adsorbed on rayon fiber at 7.5 K.

E. PEROXIDASELIKE OXIDATION

Peroxidases, widely distributed in nature, are heme enzymes that catalyze the oxidation of proton donors AH_2 with H_2O_2 as in Eq. (12). It is known that divalent phenols, aminophenol, p-aminobenzoic acid, p-phenylenediamine, and vitamin C as well as ferrocytochrome c are oxidized [8].

$$AH_2 + 2H_2O_2 \xrightarrow{\text{peroxidase}} A + 2H_2O \qquad (12)$$

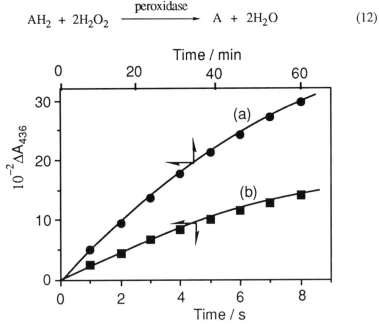

Figure 4 Oxidation of guaiacol by **6** in the presence and absence of H_2O_2 at pH 10 and 25°C. [**6**] = 5 x 10^{-5} mol/l, [guaiacol] = 2.4 x 10^{-3} mol/l. (a) Oxygen oxidation; (b) H_2O_2 oxidation, [H_2O_2] = 8.8 x 10^{-3} mol/l.

Peroxidaselike oxidation of guaiacol was studied by **6** and **6**-polyelectrolyte in the presence of H_2O_2 [9]. The results of oxidation with O_2 or H_2O_2 by **6** are shown in Fig. 4. The maximum rate of oxidation was about 1000 times greater than in the absence of H_2O_2. The final product was **10**, which is identical to the product using horseradish peroxidase. The kinetic investigation indicates that the reaction can be regarded as an ordered two-substrate reaction and that the oxidation proceeds according to the peroxidaselike mechanism in Fig. 5. The initial rate law can be expressed in Eq. (13), where V_0 is the initial rate, V_{max} is the maximum rate for oxidation of guaiacol, K_m is the Michaelis constant for H_2O_2, $K_m{}^G$ is that for guaiacol, and $K_{G\ H_2O_2}$ is a complex constant defined by Alberty [10]. It was found that V_{max} was 0.19 s^{-1}, K_m is 2.7 x 10^{-2} M, K_m is 8 x 10^{-3} M, and $K_{G\ H_2O_2}$ = 8.5 x 10^{-5} M^2.

$$\frac{V_{max}}{V_0} = 1 + \frac{K_m^G}{[Guaiacol]} + \frac{K_m^{H_2O_2}}{[H_2O_2]} + \frac{K_{G\ H_2O_2}}{[Guaiacol]\ [H_2O_2]} \tag{13}$$

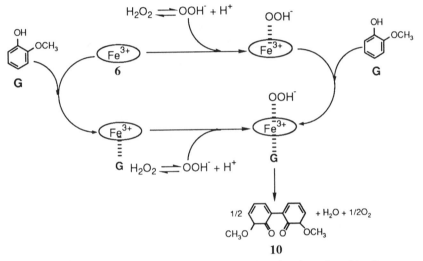

Figure 5 Reaction scheme for oxidation of guaiacol catalyzed by **6**. G, guaiacol.

Figure 6 shows the dependence of the initial rate on the *p*H of the guaiacol oxidation by **6**. V_0 increased with increasing *p*H from 8 to 10, and then drastically decreased after passing a maximum reached at *p*H 10. This behavior is attributable to the dissociation of guaiacol, whose pK_a is 10.5.

Peroxidases are heme enzymes with Fe atoms attached to protein and coordinated to protoporphyrin; that is, they are polymer-metal complexes. To perform peroxidaselike catalysis, polymer-metal complex was prepared by mixing **6** and polyelectrolyte. The oxidation of guaiacol by H_2O_2 using **6** was carried out in the presence of polyelectrolyte such as poly(vinylamine) **11** and poly(4-vinylimidazole) **12**. Figure 6 shows the *p*H dependence on the oxidation of guaiacol by the $6/H_2O_2/11$ system, whose optimum *p*H for oxidation is 5.5.

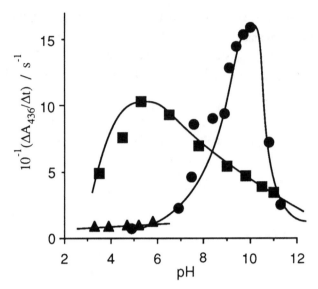

Figure 6 Effect of pH on the initial rate of guaiacol by 6/H$_2$O$_2$/polyelectrolyte systems. [H$_2$O$_2$] = 8.8 x 10^{-3} mol/l, [guaiacol] = 2.4 x 10^{-3} mol/l, [6] = 5 x 10^{-5} mol/l, [11]$_{unit}$= 1 x 10^{-2} mol/l; (■): 6/H$_2$O$_2$/11 system, (▲): 6/H$_2$O$_2$/12 system (●): 6/H$_2$O$_2$ system.

The surroundings of polymer-metal complex for 6/H$_2$O$_2$/11 system, which depend on *p*H, are illustrated in Fig. 7. At *p*H<6, a salt (-COO$^-$····-NH$_3^+$) is formed by the neutralization of the carboxy group of **6** and the amino group of **11** [Fig. 7(a)]. Consequently, the access of OOH$^-$ to the active site is accelerated by an environmental effect of positively charged polyelectrolyte. On the other hand, at *p*H>6, the free amino groups of **11** can coordinate to **6** from axial coordination sites, as illustrated in Fig. 7(b). Therefore, the free amino groups inhibit the catalysis, since the amino ligand is in competition with the OOH$^-$ ion as ligand.

The *p*H dependence of catalytic activity for the 6/H$_2$O$_2$/12 system is also shown in Fig. 6. Polyelectrolyte **12** inhibits the catalysis because of formation of stable complex by coordination of the imidazoyl group to **6**.

All the peroxidaselike reactions described above were inhibited by adding cyanide. It is thought that this phenomenon is identical to the competitive inhibition of peroxidase by cyanide ion.

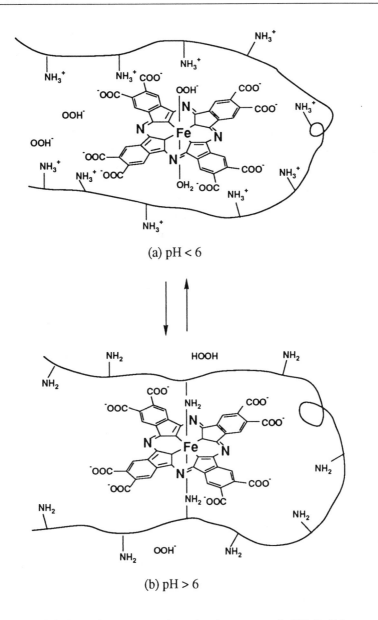

(a) pH < 6

(b) pH > 6

Figure 7 Schematic representation of active center of **6/H₂O₂/11** system for peroxidaselike oxidation.

F. APPICATION OF ENZYME-LIKE OXIDATION

Pollutants of the ecosystem are usually a mixture of volatile components generated in the decomposition by bacteria of biomaterials as proteins, carbohydrates, and higher fatty acids. About 400 kinds of offensive molecules have been identified so far, but most of the offensive smells are attributable to hydrogen sulfide, mercaptan, aldehyde, ammonia, amine, indole, and skatole. Actually, oxidative enzymes *in vivo* almost decompose these molecules, as shown in Table 4. These reactions *in vivo* suggest that the metallophthalocyanine derivatives mentioned in Sections B-E can decompose catalytically the offensive smells without ammonia and amine. Meanwhile, ammonia and amine can be removed by neturalization according to Eq. (14). With these ideas in mind, a remover of offensive substances, that is, a deodorant, has been developed using phthalocyanine derivatives [11-13].

Table 4 Enzymelike Reaction for Decomposition of Typical Offensive Molecules

Oxidaselike reaction

$$2\,RCHO + O_2 \longrightarrow RCOCOR + H_2O_2$$

$$RCH_2NH_2 + O_2 + H_2O \longrightarrow RCHO + NH_3 + H_2O_2$$

$$4\,RSH + O_2 \longrightarrow 2\,RSSR + 2\,H_2O_2$$

$$H_2SO_3 + O_2 + H_2O \longrightarrow H_2SO_4 + H_2O_2$$

Peroxidaselike reaction

Oxygenaselike reaction

$$\text{—COOH} \quad + \quad RNH_2 \quad \longrightarrow \quad \text{—COO}^- \cdots \overset{+}{N}H_3R \qquad (14)$$

Compound **6** superior to enzymelike oxidation is held to suitable supports followed by;

The formation of salt by neutralization of anionic **6** with cationic supports;
The formation of charge-transfer complexes between **6** and supports;
The van der Waals interaction between hydrophobic phthalocyanine ring and supports;
The formation covalent bond by chemical reaction **6** with supports.

The **6** adsorbed on rayon fiber mentioned in Section D, which contains **6** at 4.58 wt %, is prepared using the van der Waals interaction. Its deodorant action was studied in *in vitro* experiments. The deodorant effect of 70 mg of **6** adsorbed on rayon fiber against a mixed gas of methyl mercaptan (230 ppm), ammonia (250 ppm), and trimethylamine (200 ppm) in a volume of 1.5 l of a soft plastic bag is shown in Fig. 8. It was found that these offensive gases were almost decomposed in the plastic bag containing **6** adsorbed on rayon fiber within 1 h [11].

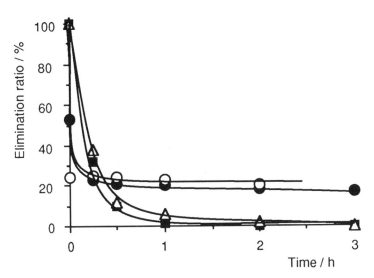

Figure 8 Elimination of bad-smelling gaseous substances by **6** adsorbed rayon fiber. (■): H_2O; (▲): CH_3SH; (●): NH_3; (○): $N(CH_3)_3$.

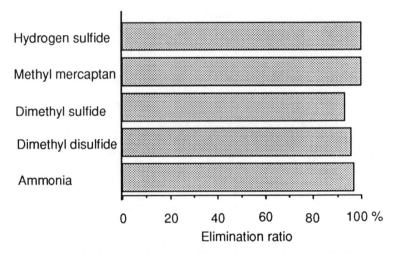

Figure 9 Elimination of bad-smelling substances in air by **6** adsorbed on rayon fiber.

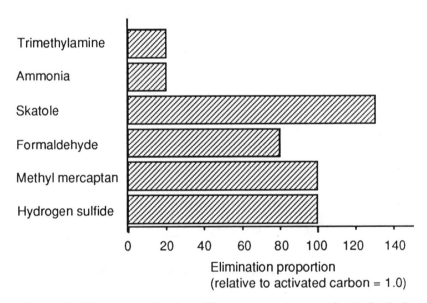

Figure 10 Elimination of bad-smelling gaseous substances by **6** adsorbed rayon fiber compared with that by activated carbon.

Air surrounding a foul-smelling dumping ground was collected in a soft plastic bag, where ammonia (0.3 ppm), dimethyl disulfide (13.5 ppb), dimethyl sulfide (2.5 ppb), methyl mercaptan (21.1 ppb), and hydrogen sulfide (43 ppb) were contained. Two grams of **6** adsorbed on rayon fiber was placed in a glass tube (20 x 110 mm^2) and 20 ml of the above air was passed through the glass tube. The passed-on air was analyzed to determine the extent to which the fiber eliminated the offensive constituents. The elimination ratios are illustrated in Fig. 9, which indicates that methyl mercaptan and hydrogen sulfide were completely eliminated by passing through the plastic tube filled with **6** adsorbed on rayon fiber and more than 90% of the ammonia, dimethyl disulfide, and dimethyl sulfide was eliminated [12].

A mixed gas of trimethylamine (800 ppm), ammonia (800 ppm), skatole (800 ppm), formaldehyde (800 ppm), methyl mercaptan (500 ppm), and hydrogen sulfide (500 ppm) was prepared and the deodorization of **6** adsorbed on rayon fiber was examined using a similar procedure as described [12]. The results as compared with activated carbon are shown in Fig. 10. It seems that **6** adsorbed on rayon fiber eliminated 20 to 30 times more of the offensive substances than did activated carbon.

For the present, practical applications of **6** are envisioned for domestic use and medical facilities [11, 13].

REFERENCES

1. L. Trynda, *Inorg. Chim. Acta,* 78 (1983) 229.
2. O. Hirabaru, T. Nakase, K. Hanabusa, H. Shirai, N. Hojo, and K. Takemoto, *J. Chem. Soc., Dalton Trans.* (1984) 1485.
3. T. Tsuiki, K. Yanagisawa, E. Masuda, T. Koyama, K. Hanabusa, and H. Shirai, *J. Phys. Chem.,* 95 (1991) 417.
4. H. Shirai, A. Maruyama, J. Takano, K. Kobayashi, N. Hojo, and K. Urushido, *Makromol. Chem.,* 181 (1980) 565.
5. H. Shirai, A. Maruyama, K. Kobayashi, and N. Hojo, *J. Polym. Sci., Polym. Lett. Ed.,* 17 (1979) 661.
6. H. Shirai, A. Maruyama, K. Kobayashi, N. Hojo, and K. Urushido, *Makromol. Chem.,* 181 (1980) 575.

7. H. Shirai, T. Tsuiki, E. Masuda, T. Koyama, and K. Hanabusa, Seminar on Macromolecule-Metal Complexes, University of Tokyo, Preprints (1987) 43.

8. B. C. Saunders, *Inorganic Biochemistry*, Vol. 2, p. 989, Elsevier, G. L. Eichhorn, Ed., Amsterdam, 1973.

9. H. Shirai, A, Maruyama, M. Konishi, and N. Hojo, *Makromol. Chem.*, 181 (1980) 1003.

10. R. A. Alberty, *J. Am. Chem. Soc.*, 75 (1953) 1928.

11. H. Shirai, *Fragrance J.*, 86 (1987) 75.

12. J. Fukui, Y. Sakai, K. Hosaka, T. Yamashita, A. Ogawa, and H. Shirai, *J. Am. Geriat. Soc.*, 90 (1990) 889.

13. H. Shirai, *J. Textile Machinary Soc.*, 40 (1987) 125.

6

Phthalocyanine

Based Liquid

Crystals: Towards

Submicronic Devices

Jacques Simon and
Pierre Bassoul

A. INTRODUCTION

Le Beau est fait d'un élément éternel, invariable, dont la quantité est excessivement difficile à déterminer, et d'un élement relatif, circonstanciel, qui sera, si l'on veut, tour à tour ou tout ensemble, l'époque, la mode, la morale, la passion.   (C. Baudelaire)

Molecular materials are made from molecular units that can be individually synthesized and characterized and that are, in a second step, organized into some condensed phase: crystal, solid, and liquid crystalline thin films [1, 2]. The chemist is faced with the problem of predicting macroscopic physical properties of materials from experimentally accessible parameters associated with the molecular unit: absorption spectra, redox potentials, magnetic properties, etc. Accurate calculations are often difficult due to the number of parameters that should be taken into account. A chemical version of the Feynman method is probably more efficient: "guess the exact solution and try afterwards to rigorously demonstrate it is really an exact solution" [3]. In the following sections this method will be extensively — while, one hopes, not abusively — used.

B. PHTHALOCYANINE DERIVATIVES AS MOLECULAR UNITS

Phthalocyanine derivatives offer a very wide choice of molecular physicochemical properties by changing the type of metal complexes and/or the nature of the substituents.

Small divalent ion complexes of unsubstituted phthalocyanine (Pc) [Cu(II), Ni(II), Zn(II), Mg(II),...] are in most cases of D_{4h} symmetry; larger ions [Pb(II), Sn(II),...] lead to out of plane complexes of lower symmetries. Judicious additional substitution of the macrocycle permits one to obtain any desired symmetry properties of the molecular unit (Fig.1).

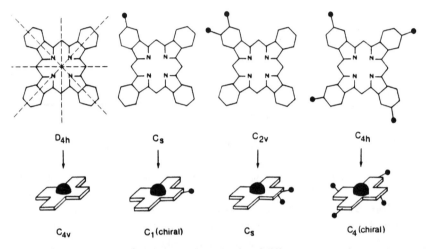

Figure 1 Some examples of phthalocyanine subunits of different symmetries. In-plane and out-of-plane complexes have been figured.

Simple chemical means permit the mastery of the symmetry of the molecular unit. The physicochemical properties of the corresponding condensed phases will be, however, difficult to deduce from the characteristics of the molecular units. It is possible to relate the symmetry of the condensed phases to the symmetry of the molecular unit [4]. The molecular symmetry elements play a role only when they are in common with the symmetry elements of the condensed phase (site symmetry). The reunion of several highly unsymmetrical molecular units may lead to highly symmetrical unit cells, whereas highly symmetrical molecular units may lead to low site symmetries [5]. However, two principles seem to be obeyed: (1) the most closely packed material is always favored; (2) for the same compacity, the highest site symmetry is preferred [5]. The shape of the molecules imposes local spatial limitations. It is therefore necessary to define the notion of molecular shape [5–8] and to determine the number of ways they can be arranged to form dense materials. This is a discouragingly difficult task for three-dimensional single crystals.

The establishment of a relationship between the molecular unit characteristics and the type of organization seems less difficult for mesophases. Mesophases and liquid crystals are characterized by the presence of structural disorders in some spatial directions. The mesophases can be classified depending on the type of disorder (translational or rotational) present (Fig. 2).

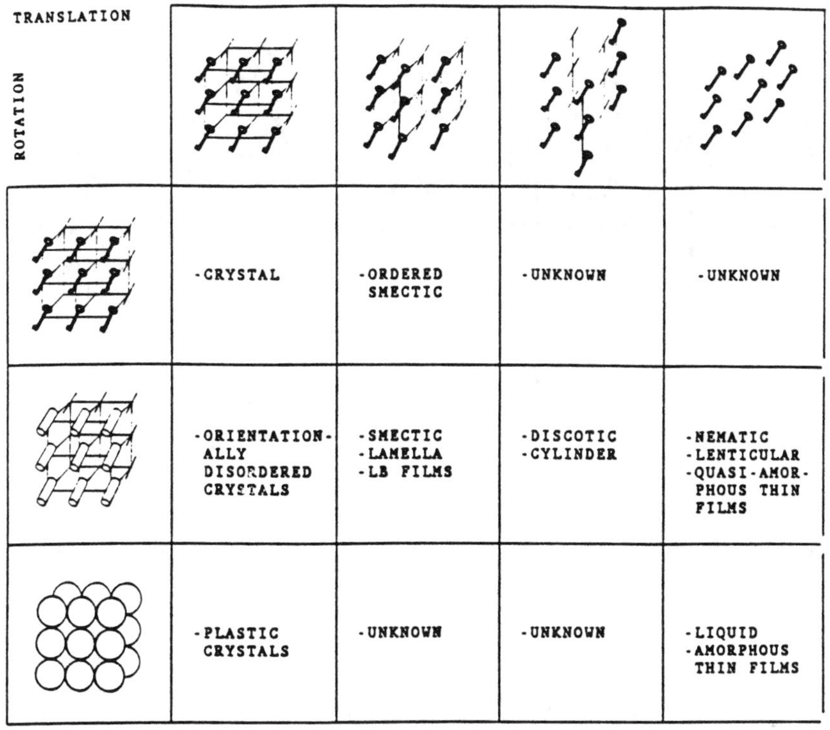

TRANSLATION				
	·CRYSTAL	·ORDERED SMECTIC	·UNKNOWN	·UNKNOWN
	·ORIENTATION- ALLY DISORDERED CRYSTALS	·SMECTIC ·LAMELLA ·LB FILMS	·DISCOTIC ·CYLINDER	·NEMATIC ·LENTICULAR ·QUASI-AMOR- PHOUS THIN FILMS
	·PLASTIC CRYSTALS	·UNKNOWN	·UNKNOWN	·LIQUID ·AMORPHOUS THIN FILMS

Figure 2 Classification of mesophases according to rotational or translational order breakings (after [4]).

In this case, two-dimensional (^{2}D) arrangements are very commonly encountered. The relationship between ^{2}D molecular shapes and ^{2}D packings is therefore of importance [4, 6]. The ^{2}D shape of a molecule may be approximated by elementary forms (squares, triangles, circles) whose number is related to the complexity of the molecule. These elementary shapes are then organized in ^{2}D in order to follow the closest-packing principle. [4, 9–11]. Depending on the substituents, various elementary forms may be found for phthalocyanines (Fig. 3).

Most of the liquid crystals based on Pc or porphyrins contain alkyl side chains [12–16, 167]. In this case, the molecular units may be approximated by cylinders (Fig. 3).

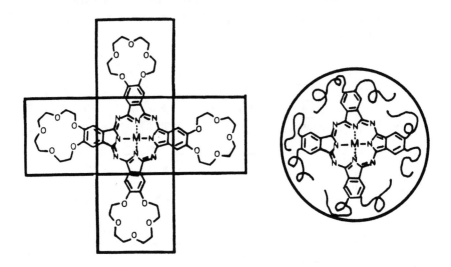

Figure 3 Two models for the shape of substituted phthalocyanine derivatives.

A more accurate and precise way of defining the shape of the molecular units may be established from topological arguments [9–11]. This is the object of the following section.

C. FACTORS INFLUENCING THE PACKING IN MESOPHASES

Chemists are faced with the problem of estimating the structure and the type of packing of materials knowing the nature and the characteristics of the constitutive molecular units. It is in principle possible to determine the most stable packing by modeling the intermolecular interactions with suitable potentials. However, in practice, this procedure involves prohibitively long computational times. In this section it is proposed to determine qualitatively the parameters that are relevant for determining the arrangement of molecular units in mesophases. It is worth emphasizing that the approach taken does not pretend to predict accurately the type of packing. One only hopes to provide clues to chemists for finding new molecular units and new types of arrangements.

i. Shape of the Molecular Units

It is generally accepted that conventional liquid crystals — nematic, smectic, discotic, plastic crystals — are derived from mesogens whose shape may be approximated by cigars, disks, or spheres. The molecule itself can hardly be seen as a cigar or a disk. However, in the mesophases, rapid conformational changes occur, and time-averaged conformations may yield the model shape.

The closest-packing criterion [5] is the most reliable way to predict the molecular arrangements in condensed phases. In general [5], the closest ^{2}D packing criterion is postulated to be fulfilled whenever the coordination number of the molecular unit is 6 or more. Arbitrarily shaped molecules give rise to limited possibilities of site symmetries and space groups [5]. For the same compactness, the highest site symmetry is always favored. An alternative way relying on a ^{2}D tiling approach may be attempted.

It has been demonstrated that a finite number of isohedral elementary tiles can be found that cover a plane without gap or overlap [9]. The elements of an isohedral tiling are related by symmetry operations of its periodic ^{2}D space group: All the elementary tiles are equivalent. Topological arguments permit one to define 11 different elementary tiles (Laves tiles). Each tile is defined by a number of " singular points," which are the vertices common to at least three different tiles. The topological classes are conserved by stretching or compressing the plane drawing without folding or tearing it.

The symmetry properties of the tilings are described by one of the 17 ^{2}D space groups and by the site symmetry of the tile. By combining symmetry and topology, 93 types of isohedral tiling, denoted IH1 to IH93 [9] are defined. These may be considered as the elementary molecular shapes [11]. Any molecular unit may be related to an elementary shape that is chosen to ensure the maximum compactness. The case of the crown-ether-substituted phthalocyanine will be more thoroughly treated as an example.

The symmetry of the molecule is D_{4h}; two elementary shapes may be found that preserve a d_4(IH76: $1C_4$, $2\sigma_v$) or a C_4(IH62: $1C_4$) site symmetry. In the elementary shape IH76, no deformation is possible, and only the position of the molecule relative to the singular points may be varied (Fig. 4).

The compactness of the ^2D lattice is given by the ratio of the molecular area to the surface of the elementary shape.

The elementary shape IH62 may be deformed in some extent (the C_4 symmetry is conserved in all cases). The best compactness is obtained when the area of the elementary shape is as small as possible. (See Fig. 5)

However, since the elementary shape must tile the plane, the parts of the molecule outside the elementary polygonal tile must correspond to relative concavities (yin–yang principle [11]).

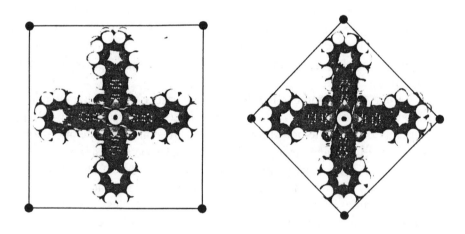

Figure 4 The fitting of the (15-crown-5)₄Pc molecular unit with the elementary tile IH76.

The relationship that can be found between the shape of a molecular unit and the type of packing has been briefly explored in the previous section. Another parameter is important for determining the type of organization: the chemical nature of the moieties constituting the molecular unit. "Like as like goes together." Chemistry is not original in this respect: in condensed phases, segregation in microdomains will occur in order to respect the affinities of the various chemical fragments of the molecule (sympathy–antipathy principle).

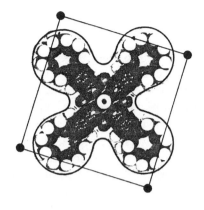

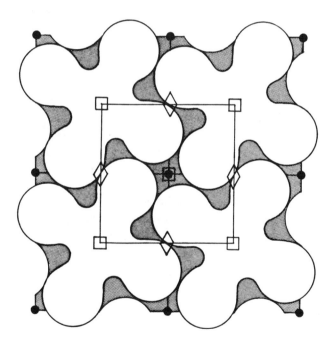

Figure 5 The elementary shape IH62 associated with the (15-crown-5)₄Pc molecular unit. The corresponding lattice and its symmetry elements are figured.

ii . Segregation in Mesophases:
Antipathy – Sympathy Principle

The segregation of molecular fragments to form chemically homogeneous microdomains is very often encountered: Rigid aromatic subunits pile up with each other; paraffinic chains do not mix with perfluoroalkyl chains; lipophilic groups separate from hydrophobic moieties; etc. [17, 18]. This effect may be stated as a general sympathy–antipathy principle. However, a more precise insight into this phenomenon may be gained from simple physical principles.

Let us consider N_i moles of the compound i. In the condensed phase each molecule is surrounded by Z other molecular units [19]. Z is considered to be unchanged with the type of molecules (regular solution case). If E_{ii} is the energy of interaction between two molecules i, the total energy within the condensed phase is for pure phases:

$$Z\frac{N_i}{2}E_{ii} \tag{1}$$

In the case in which, two different molecules (1and 2) are considered, each molecule 1 is surrounded by $ZN_1/(N_1+N_2)$ molecules of 1 and $ZN_2/(N_1+N_2)$ molecules of 2; then the total energy of interaction is :

$$\frac{1}{2}Z\frac{(N_1)^2}{N_1+N_2}E_{11}+Z\frac{N_1N_2}{N_1+N_2}E_{12}+\frac{1}{2}Z\frac{(N_2)^2}{N_1+N_2}E_{22} \tag{2}$$

The difference of the energy between the segregated phases and the mixture is the mixing enthalpy ΔH_M and is given by subtracting Eq. (2) from Eq. (1) :

$$\Delta H_M = Z\frac{N_1N_2}{N_1+N_2}\Delta_{12} \tag{3}$$

with

$$\Delta_{12} = \frac{E_{11}}{2} + \frac{E_{22}}{2} - E_{12} \tag{4}$$

It is important to stress that Δ_{12} is always positive if specific interactions, such as hydrogen bonds, are excluded. The nonbonded interactions between molecular units can be expressed by the equation :

$$E_{ij} \approx 1/(r_{ij})^n \tag{5}$$

where r_{ij} is the distance between the molecular units i and j and n is a parameter that depends on the type of electrostatic interactions considered; Δ_{12} is then given by :

$$\Delta_{12} \approx \frac{1}{(2r_1)^n} + \frac{1}{(2r_2)^n} - \frac{2}{(r_1+r_2)^n} \tag{6}$$

It can be readily seen that $\Delta_{12}=0$ if $r_1=r_2$ and that Δ_{12} is always positive if $r_1 \neq r_2$. The enthalpy of mixing is therefore always positive and unfavorable (Fig. 6).

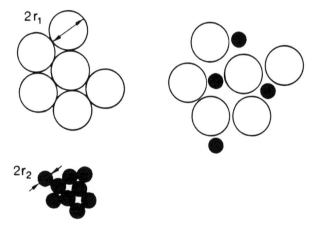

Figure 6 Segregated domains and mixtures of two molecular units of different size.

The overall tendency to segregate is given by the free energy of mixing ΔG_M:

$$\Delta G_M = \Delta H_M - T \Delta S_M$$

ΔS_M is always positive (the disorder increases by mixing), and segregation will occur whenever ΔH_M is larger than $T \Delta S_M$. From the equation for Δ_{12}

it is clear that fragments of the same chemical nature (and therefore the same size) will tend to form homogeneous microdomains. A more quantitative estimation of the reciprocal "antipathies" of chemically different fragments may be gained from the solubility parameter values established for polymers [19, 20].

When apolar molecular units are considered, the main contribution to the interaction energy arises from van der Waals interactions [19]:

$$E_{ij} = -\frac{3}{2} \frac{\alpha_i \alpha_j}{(d_{ij})^6} \frac{l_i l_j}{l_i + l_j} \tag{7}$$

where $d_{ij} = r_i + r_j$; I is the ionization potential; and α_i is the polarizability coefficient. With such a potential, in the case the two molecular units have not too different physicochemical properties:

$$4 I_1 I_2 \approx (I_1 + I_2)^2 \tag{8}$$

$$(4 r_1 r_2) \approx (r_1 + r_2)^2 \tag{9}$$

Under these conditions:

$$E_{12} \approx - \sqrt{E_{11} E_{22}} \tag{10}$$

Equation (3) may then be written:

$$\Delta H_M = \frac{N_1 N_2}{N_1 + N_2} \left[\left(\frac{E_{11} Z}{2} \right)^{1/2} - \left(\frac{E_{22} Z}{2} \right)^{1/2} \right]^2 \tag{11}$$

When the molecular units have very different sizes, this expression becomes:

$$\Delta H_M = V_M \left[\left(\frac{\Delta E_1}{V_1} \right)^{1/2} - \left(\frac{\Delta E_2}{V_2} \right)^{1/2} \right]^2 \Phi_1 \Phi_2 \tag{12}$$

where V_M is the total volume of the mixture; Φ_i the volume fraction of component 1 or 2 in the mixture; $\Phi i = \frac{N_i V_i}{V_M}$; N_i the number of moles of species i; and V_i the partial molar volume.

The quantity $(\Delta E/V)^{1/2}$ is often referred to as the solubility parameter δ; $\Delta E/V$ is the cohesive energy:

$$\delta_i = \left(\frac{\Delta E_i}{V_i}\right)^{1/2} \tag{13}$$

Since the entropy ΔS_M is always positive, mixing will occur ($\Delta G_M < 0$) whenever ΔH_M is not too large; consequently $(\delta_1 - \delta_2)^2$ must be relatively small. The solubility parameter δ of any molecule or molecular subunit may be estimated from their formula [20]. To each chemical group is associated a "molar-attraction constant" K; the solubility parameter is obtained by summing K over all the atoms and groupings in the molecules :

$$\delta = d \frac{\Sigma K}{M_W} \tag{14}$$

where d is the density and M_W the molecular weight

The values of the group molar attraction constants K are derived from the measurements of heats of vaporization [21] or from vapor pressure determinations [22].

The use of group molar attraction constants is of considerable interest for predicting the type of mesophases that can be obtained (Table 1) The molecular unit is first divided into subunits of homogeneous chemical nature. The solubility parameters have been determined for high molecular weight polymers, and, therefore, fragments of sufficient size must be considered. In the case of substituted phthalocyanines, the molecular subunits will be the rigid aromatic core and the side chains (Fig. 7).

Table 1 Group Molar Attraction Constants Derived from the Measurements of Vapor Pressures.

Group		K	Group		K
-CH$_3$		147.3	-OH		225.84
-CH$_2$-		131.5	-H	acidic dimer	-50.47
>CH		85.99	OH	aromatic	170.99
>C<		32.03	NH$_2$		226.56
			-NH-		180.03
CH$_2$=		126.54	-N-		61.08
-CH=		121.53	-C≡N		354.56
>C=		84.51	NCO		358.66
-CH=	aromatic	117.12	-S-		209.42
-C=	aromatic	98.12	Cl$_2$		342.67
-O-	ether,acetal	114.98	Cl	primary	205.06
-O-	epoxide	176.20	Cl	secondary	208.27
-COO-		326.58	Cl	aromatic	161.0
>C=O		262.96	Br		257.88
-CHO		292.64	Br	aromatic	205.60
(CO)$_2$O		567.29	F		41.33

After [20, 22] The solubility parameters calculated from K are in cal$^{1/2}$ cm$^{-3/2}$.

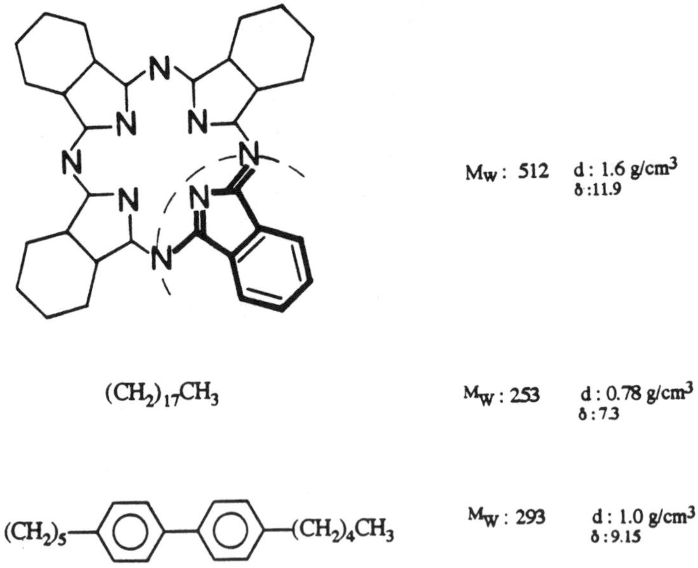

M_w: 512   d: 1.6 g/cm^3
               δ :11.9

(CH$_2$)$_{17}$CH$_3$

M_w: 253   d: 0.78 g/cm^3
               δ : 7.3

(CH$_2$)$_5$—〇—〇—(CH$_2$)$_4$CH$_3$

M_w: 293   d: 1.0 g/cm^3
               δ : 9.15

Figure 7 Calculation of the solubility parameters (antipathy factor) of alkyl-substituted phthalocyanine fragments.

It is apparent that strong "antipathy" occurs between the phthalocyanine rigid core and the paraffinic chains. The solubility parameter δ for linear paraffinic chains is almost independent of the molecular weight if chain ends are neglected. The segregation is however reinforced with high molecular weight fragments because of the decrease of the mixing entropy :

$$\Delta S_M = - RV_M\left(\frac{\Phi_1}{V_1}\ln\ \Phi_1 + \frac{\Phi_2}{V_2}\ln\ \Phi_2\right) \qquad (15)$$

In the case of phthalocyanine macrocycles substituted with long alkyl chains [12, 13], the segregation that must occur between the rigid aromatic moieties and the flexible alkyl chains leads to columnar mesophases (Fig. 8).

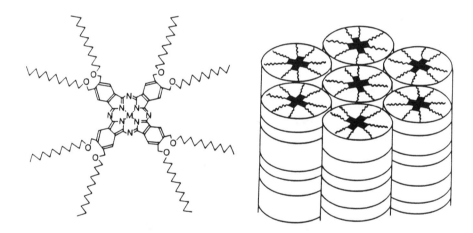

Figure 8 Columnar liquid crystals formed from octaalkoxymethyl–substituted phthalocyanines (after [12]).

iii. Geometrical Parameters Associated with Paraffinic Chains

Models established for polymers can also be useful for estimating the geometrical parameters associated with side chains (see [20]).

The mean-square end-to-end distance of a linear chain molecule is usually expressed as :

$$<r^2> = <r^2>_0 \alpha \qquad (16)$$

where $<r^2>_0$ is the unperturbed end-to-end distance and α is the factor including long-range interactions. $<r^2>_0$ represents the dimension of the chain taking into account bond-angle restrictions and steric hindrance to internal rotation. The quantity α is associated with "long-range interactions" as the swelling of the chain by solvent–polymer interactions; it will be ignored in the following discussion.

A hypothetical freely rotating state may be postulated in which bond-angle restrictions are retained but steric hindrances to rotation are released. The mean-square end to-end-distance of the freely rotating chains $<r^2>_{of}$ may be readily calculated from the structure of the chain [20] :

$$<r^2>_{of} = n_b l_b^2 \left(\frac{1+\cos\,\theta}{1-\cos\,\theta} \right) \qquad (17)$$

where n_b is the number of bonds, θ the supplement of the valence bond angle, and l_b the bond length

The ratio of $<r^2>_0$ to $<r^2>_{of}$ is a measure of the effect of steric hindrance on the average chain dimension (Table 2). This ratio is independent of the molecular weight.

Polysiloxane and polyethyleneoxide chains are less affected by steric hindrance, while polyacrylamide and polyvinylcarbazole have fairly rigid backbones.

The end-to-end distance estimations (see Table 3) are only valid for polymeric chains. In the case of short chains, the stiffness considerably increases, and end chains can no longer be ignored. In this case a Flory type calculation may be carried out [17, 23]. For paraffinic chains, three characteristic distances must be considered: the value corresponding to the fully elongated chain (l_{ext}) as present in crystalline materials, the hypothetical freely rotating state (l_{free}), and various intermediate conformations in which steric hindrance is taken into account (Flory calculations or σ value).

Table 2 Values of $\sigma = \langle r^2 \rangle_0 / \langle r^2 \rangle_{of}$ for Various Polymeric Chains [a]

Structure	σ	Structure	σ
(propene-like)	1.7	$-CH_2-CH$ (carbazolyl)	2.8
(1-butene-like)	1.2		
$-CH_2-$	1.6-1.8	OCH_2CH_2O	1.4
$-CH_2-CH$, $C=O$, NH_2	2.7	$-Si-O-$ with CH_3 , CH_3	1.3
$-CH_2-CH$ (phenyl)	2.2		

[a] After [20].

Table 3 End-to-End Distances for Paraffinic Chains as a Function of the Number of $-CH_2-$ Subunits (n) ·

n	l_{ext}	l_{free}	l_{flex}	$l_{free}\sigma$
4	5.1	4.3	4.6	—
6	7.6	5.3	6.6	—
8	10.1	6.1	8.4	—
10	12.6	6.8	9.9	—
12	15.2	7.5	11.4	12.7
18	22.8	9.2	14.4	15.6

l_{ext}: fully elongated conformation ($l_{ext} = 1.265n$); l_{free}: freely rotating state ($l_{free} = \sqrt{4.66n}$; l_{flex} = flexible conformation; $l_{flex} = pl_{ext}$ (p is obtained from a Flory model, see [23]); $\sigma = \langle r^2 \rangle_0 / \langle r^2 \rangle_{of}$. The overall chain length is obtained by adding end chain parameters when not negligible (from [1, 20]).

iv. Geometrical Constraints on Packing

Some molecular restrictions on packing have been explored: the molecular shape, the antipathy character, the geometries associated with more or less flexible side chains, etc. It is now possible to use these notions to predict some relationships between the molecular unit characteristics and the type of organization of the mesophase. The problem of estimating the number of alkyl chains necessary to form hexagonal columnar liquid crystals may be taken as an example.

The phthalocyanine subunit may be approximated by a St. André cross consisting of five squares (5×5 Å). The end-to-end distance of the paraffinic side chains cannot be the less than the free rotating value, that is, 9.2 Å for the octadecyl chains (Fig. 9).

The thickness of the rigid core is the van der Waals value for aromatic hydrocarbons (3.4 Å). A circular shape is needed to form a hexagonal array; the minimum volume of the molecular unit is therefore $\pi d^2 h/4$, with $d = 15+2(9.2+2.1)$ and $h = 3.4$; $V = 3773$ Å^3. The elementary volume of an alkyl chain may be estimated from its area (~20 Å^2 [24]) and its length: $(22.8+2.1)20 = 498$ Å^3. The minimum volume of the cylinder defined by the size of the phthalocyanine ring (425 Å^3) and the end-to-end distance of the side chains is 2979 Å^3; it is therefore necessary to have: $(3773-425)/498 = 6.7$ alkyl chains to occupy the elementary cylinder fully. This prediction seems to agree with experimental findings: Columnar mesophases are obtained with six or eight alkyl chains around the phthalocyanine ring [12, 25], whereas macrocycles substituted with only four side chains do not yield columnar liquid crystals [26, 27]. The previous calculation only holds when there is no overlap between adjacent columns.

Figure 9   Schematic representation of an alkylchain–substituted phthalocyanine.

The interplane and interchain van der Waals distances for aromatic compounds and aliphatic derivatives are 3.4 and 4.5 Å, respectively [28]. In an eclipsed arrangement of the Pc macrocycles, the packing of the alkyl chains leads to empty spaces between the aromatic moieties. To avoid these cavities, the aromatic cores may form a tilting angle θ with the paraffinic chains such that $\cos \theta = (3.4/4.5)$ ($\theta \approx 40°$) (Fig. 10).

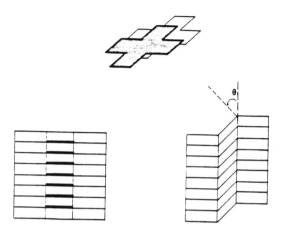

Figure 10 Two types of arrangements in columnar mesophases.

Alternatively, a staggered arrangement may be adopted. These two types of columnar mesophases have been encountered, depending on the nature of the connecting links of the side chains [12, 29–31, 168].

It is possible to explore further the few guidelines that can be found for fabricating new materials. Let us consider a disk-shaped molecular unit substituted with several chemically distinct side chains (Fig. 11). The solubility parameters of the substituents are such that crossed interactions are energetically unfavorable .

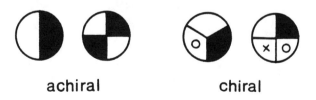

achiral   chiral

Figure 11 Molecular units substituted with chemically different substituents (antagonistic side chains).

Disk-shaped molecular units are expected to pack according to a hexagonal lattice; this latter cannot be preserved with two different types of side chains forming segregated domains (Fig.12)

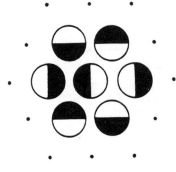

Figure 12 Illustration of the impossibility of packing molecular units possessing two chemically different substituents according to a hexagonal lattice.

The hexagonal lattice will therefore deform into lower-symmetry arrangements such as square or rectangular lattices. In the case ^{2}D chiral molecular units are used, chiral discrimination may occur, since the various possibilities of arrangements of the pure optical isomers or of the racemic mixtures are not energetically or geometrically equivalent (Fig. 13).

Theoretical considerations have been reported on ^{2}D chiral discrimination in Langmuir–Blodgett thin films [32].

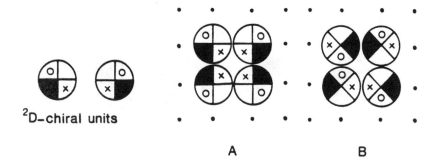

Figure 13 ^{2}D chiral molecular units and representation of the two possible arrangements within a square lattice. (A) Racemic mixture; (B) optically pure isomer.

D. MOLECULAR ORBITAL DESCRIPTIONS OF THE MOLECULAR UNIT AND OF THE MESOPHASES

The molecular unit cannot be merely described by its shape. Many of its properties will indeed depend on the type and the symmetry of the orbitals involved in the interunit interaction.

Several calculations have been carried out to determine the orbital ordering in Pc [33–36]. In D_{4h} symmetry, the orbitals involved in the low-energy optical absorption bands are:

$$
\begin{array}{lcll}
a_{1u}(\pi) & \rightarrow & e_g(\pi) & 658\text{ nm} \\
a_{2u}(\pi) & \rightarrow & e_g(\pi) & 320\text{ nm} \\
b_{2g}(d_{xy}) & \rightarrow & b_{1u}(\pi) & 283\text{ nm}
\end{array}
$$

The symmetry of the transition moments is E_u (polarization: x,y) for the first two transitions and A_{2u} (polarization z) for the last one.

When the molecular units stack to form columns, rotation around the axis of the column is allowed, and more or less stable arrangements are obtained depending on the angle θ between the macrocycle rings (Fig. 14).

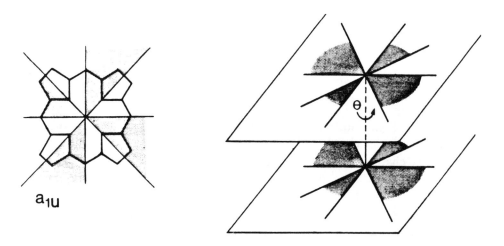

Figure 14 Symmetry of the highest occupied molecular orbital (HOMO) of phthalocyanine subunits (symmetry D_{4h}) and representation of the intermacrocyclic rotation within the columns.

The orbital overlap reaches maxima for both the fully eclipsed and the fully staggered arrangements [36] with the less favorable overlap for $\theta = 20°$.

It can now be seen how the molecular-orbital symmetries will influence physicochemical properties of the molecular unit, such as its ground-state dipole moments, its polarizability, or its hyperpolarizability.

In order to have a ground-state dipole moment, the symmetry of the molecular unit must belong to a polar symmetry group: $C_{\infty v}$ or one of its subgroups (C_n, C_{∞}, C_{nv}). The polar axis is the principal axis of symmetry of the polar group. Polar phthalocyanine molecular subunits must therefore be substituted with polar groups or transformed into out-of-plane complexes (Fig. 15).

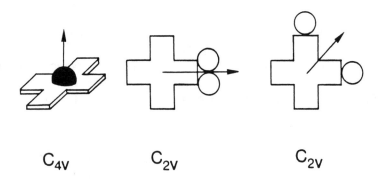

$$C_{4V} \qquad C_{2V} \qquad C_{2V}$$

Figure 15 Schematic representation of some ways of obtaining in-plane and out-of-plane polar subunits.

Such polar molecular units may be used for making liquid crystal visualization devices [37–39].

The characteristics of the optical absorption bands and of a molecular unit may be related to its polarizability [4]. Light absorption occurs when there is resonance between the ground and the excited electronic state. Transmission of light (transparency domain) involves the polarization of the electronic cloud (out-of-resonance interactions or "forced vibration" regime) [4]. The polarizability tensor coefficients may therefore be related to the optical absorption bands. The $a_{1u}(\pi) \rightarrow e_g(\pi)$ band corresponds to a transition moment of symmetry E_u. The polarization of the transition is in the (x, y) plane and is degenerate. The two polarizability tensor coefficients are therefore equal : $\alpha_{xx} = \alpha_{yy}$. A contribution along z is expected from the $b_{2g}(d_{xy}) \rightarrow b_{1u}(\pi)$ transition (symmetry: A_{2u}, polarization: z) (Fig. 16) [4].

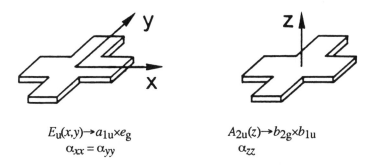

$$E_u(x,y) \rightarrow a_{1u} \times e_g \qquad\qquad A_{2u}(z) \rightarrow b_{2g} \times b_{1u}$$
$$\alpha_{xx} = \alpha_{yy} \qquad\qquad\qquad \alpha_{zz}$$

Figure 16 Symmetry of the transition moments involved in the optical absorption spectra of metallophthalocyanine and the corresponding polarizability coefficients arising from these bands.

The polarizability coefficients α may be related to the refractive indices (n) and to the birefringence of the molecular unit [in the isotropic case $\alpha \approx (n_i^2 - 1)/(n_i^2 + 2)$].

The second-order polarizability β_{ijk} may be similarly derived from the ground- and excited-state orbital symmetries. Polarization due to a two-photon interactions will occur whenever [4]:

1. The irreducible representation associated with the transition moment of the optical absorption band corresponds to the irreducible representations of x and/or y and/or z;

2. The previous irreducible representation is also associated with the representations of x^2 and/or y^2 and/or z^2 and/or xy and/or xz and/or yz.

This is exemplified in the case of the C_{2v} symmetry (Fig. 17).

Figure 17 Molecular units of symmetry C_{2v}.

The irreducible representations corresponding to C_{2v} are:

$$
\begin{array}{lll}
A_1 & z & x^2, y^2, z^2 \\
B_1 & x & xz \\
B_2 & y & yz
\end{array}
$$

The transition of A_1 symmetry (x^2, y^2, or z^2) can generate, in the C_{2v} group, a polarization along the z axis associated with the coefficients β_{zxx}, β_{zyy} and β_{zzz}, respectively. The overall molecular hyperpolarizability tensor is:

$$
\begin{array}{cccccccc}
0 & 0 & 0 & 0 & \beta_{xxz} & 0 & B_1 \\
0 & 0 & 0 & \beta_{yyz} & 0 & 0 & B_2 \\
\beta_{zxx} & \beta_{zyy} & \beta_{zzz} & 0 & 0 & 0 & A_1
\end{array}
$$

The optical absorption bands with transition moments of symmetry A_1, B_1, and B_2 will all contribute to the tensor coefficients.

The macroscopic polarizability and hyperpolarizability coefficients of materials may in turn be derived from the molecular characteristics [4, 40]. In a condensed phase, the symmetry of the molecular unit must be replaced by the site symmetry. The induced group, or site symmetry group of a molecule, is determined by the symmetry elements common to the crystal and to the molecule. When the primitive unit cell contains only one molecule, all the symmetry elements of the crystal are conserved at the molecular unit level. The number of molecules in the primitive unit cell is equal to the ratio of the order of the crystalline class upon the order of the site symmetry group. In the case of mesophases, site symmetry of higher order than molecular unit symmetry are commonly found.

The mesophases are, by definition, only partially organized, and the statistical distribution of molecules along certain directions yields highly symmetrical condensed phases. A C_{2v} molecular unit forming various liquid crystalline phases may be taken as an example (Fig. 18).

At a given time, the rotation of the C_{2v} molecular unit around the column axis of the discotic mesophases is frozen, and the local site symmetry departs from the expected D_{6h} symmetry. However, the distribution of orientations within the columns is random, and an overall D_{6h} symmetry will be detected in the x-ray pattern as long as the distribution is randomized over a distance less than the coherence length of the x-rays (200–300 Å). It may be considered that the apparent molecular unit symmetry is D_{6h} at a macroscopic level if all orientations are equally probable.

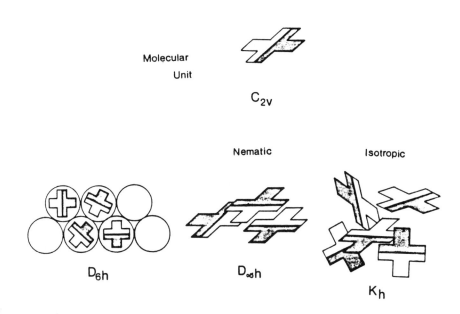

Figure 18 Formation of various mesophases derived from a C_{2v} molecular unit.

E. MESOPHASES FORMED FROM METALLO-ORGANIC COMPLEXES

Pioneer studies in the 1970s [41, 42] described the mesomorphic properties of silicon or mercury-containing compounds. Almost at the same time, the thermotropic behavior of ferrocene derivatives of Schiff's bases was reported [43–45]. Metal complexes [46, 47] or crown-ether [48–50] substituted with long alkyl chains were first described in 1977. Since then various structurally related compounds have been described [51–62] (Fig. 19). (For a recent review see [169])

Previous to our work [12], only one study concerned the phthalocyanine subunit. A tetracarboxylic derivative of metallophthalocyanine was described to form lyotropic discotic mesophases in the presence of water [63]; subsequent x-ray determinations did not confirm this assignment [64]. A porphyrin derivative substituted with long paraffinic side chains was shown to form a mesophase over 0.1°C by optical microscopy [65].

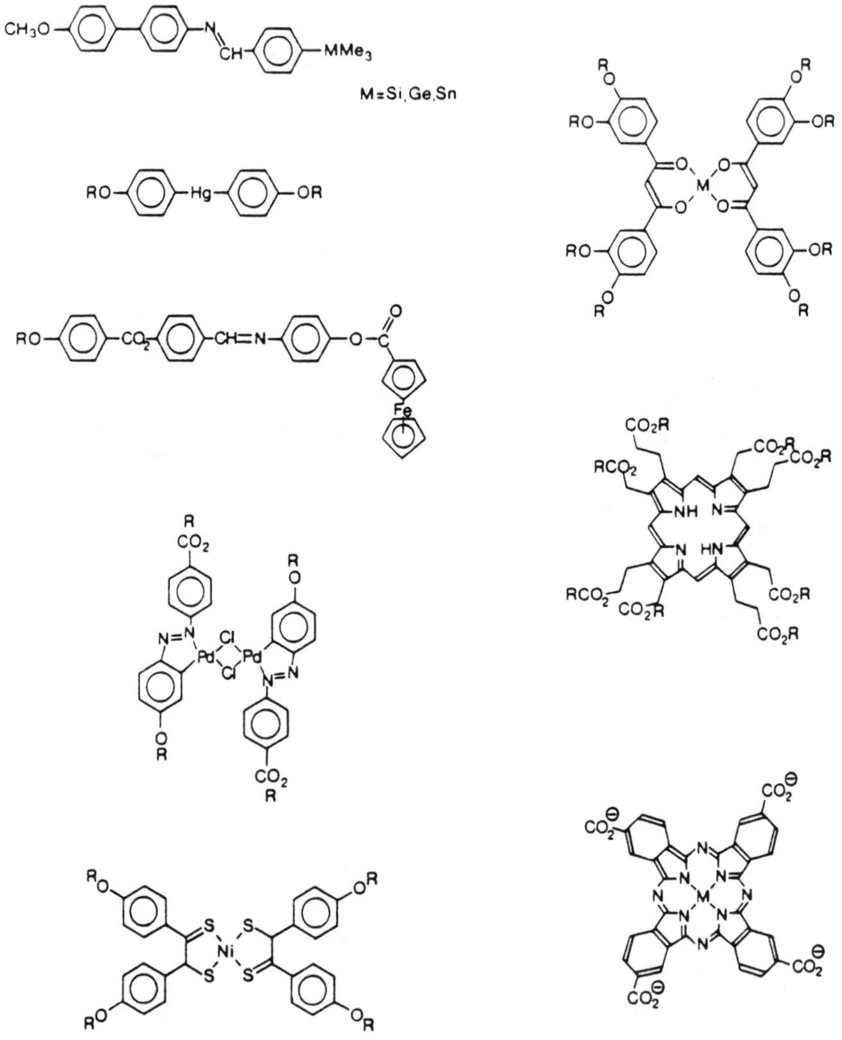

Figure 19 The main metallo-organic subunits forming mesophases.

F. SYNTHESIS AND MESOMORPHIC PROPERTIES OF PHTHALOCYANINE DERIVATIVES

i. Synthesis

Phthalocyanine macrocycles are usually synthesized by tetramerization of the corresponding 1,2-phthalonitrile or 1,2-diacide in the presence or absence of a catalyst [66, 67]. A general chemical pathway for obtaining octasubstituted phthalocyanine derivatives was first proposed by Hanack et al. [68]. Subsequently, the preparation of various octasubstituted phthalocyanine derivatives was reported with –OMe and –OC$_8$H$_{17}$ [69], –OMe [70, 71], –Me [72], and –CF$_3$ [68, 73] side groups. Three different starting materials are used (Fig. 20).

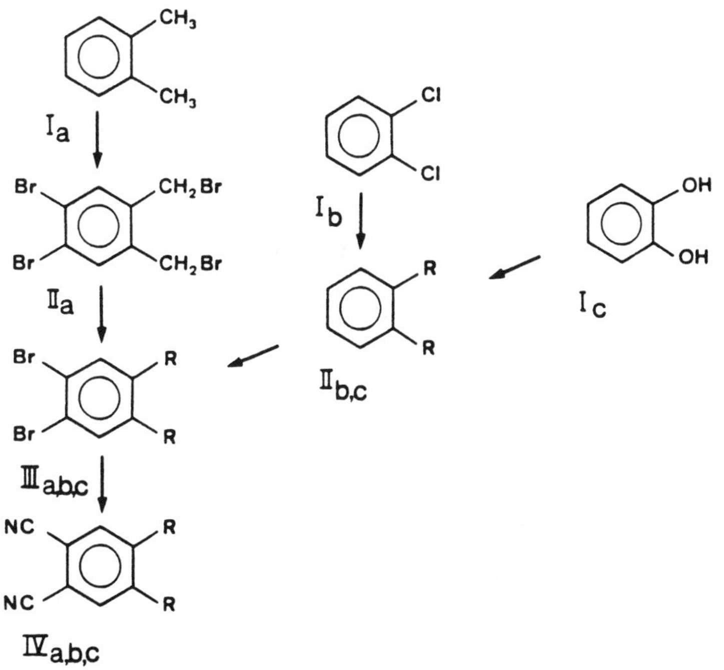

Figure 20 Chemical pathways used for synthesizing substituted phthalonitriles. (a) From [68], R = –CH$_2$OC$_n$H$_{2n+1}$, overall yield 5%; (b) from [74], R = –C$_n$H$_{2n+1}$, overall yield 6–15%; (c) from [69, 75], R = –OC$_n$H$_{2n+1}$, overall yield 60%.

In the original method *o*-xylene is treated with bromine to give the tetrabromo derivative (IIa) in approximately 25% yield; substitution with the potassium salt of a long-chain alcohol in a non-nucleophilic solvent (*t*-BuOH) [13] yields III$_a$ in 50% yield. This latter is transformed into the dicyano derivative with CuCN in DMF (yield: 40%) [12, 13].

The overall yield of this route is 5%. Higher yields are obtained by starting from catechol [74, 75] (overall yield: 60%). The alkyl side chain is in this case linked to the macrocycle via an oxygen heteroatom. It will be shown in the next sections that this introduces important changes in the mesomorphic properties of the compounds due to the difference of chain conformation induced by the two linkages [76].

The alkyl chains may also be added [74] by treating *o*-dichlorobenzene with an alkylmagnesium bromide derivative in the presence of a phosphine nickel complex [77] as catalyst. The yield for this route was not optimized and varies between 6 and 15% [74].

3,6-Dialkyl substitution instead of 4,5-disubstitution may be obtained by using a different approach [14]: 2,5-dialkylfurans are converted in a one-pot reaction into the corresponding phthalonitriles via a Diels–Alder reaction with NCCH = CHCN [14, 78]. Aromatization leads to the corresponding phthalonitrile (overall yield: 7–25%) (Fig. 21).

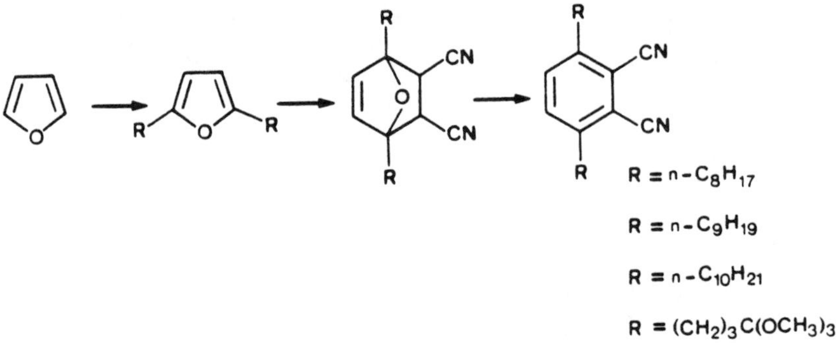

Figure 21 Chemical pathway used to synthesize 3,6-dialkyl-substituted phthalonitriles (after [78]).

Octacarboxy-substituted Pc may be obtained by esterification of the acid with long-chain alcohols [79]. The acid is synthesized in three steps starting from pyromellitic acid [80-82].

The phthalonitrile derivatives are converted into the corresponding phthalocyanines in various ways. The treatment with lithium in 1-pentanol affords the dilithium salt of Pc, which, in acidic media, leads to the metal-free derivative [83, 84]. The dicyano derivatives may be alternatively heated in high-boiling-point solvents (1-chloronaphthalene, quinoline,...) in the presence of a finely divided metal or a metallic salt to give the corresponding PcM [66, 85]. The dicyano derivatives may be first transformed into 1,3-diiminoisoindoline by reaction with NH_3 in basic media [86, 87]. The diiminoisoindolines are then dissolved in 2-dimethyl-aminoethanol and heated under reflux to give s-PcH$_2$ [87]. Most of the usual metal complexes may be readily obtained from the dianion s-Pc^{2-} obtained with a sodium or potassium alcoholate, which is reacted with the suitable metallic salt. This reaction is generally quantitative if a solvent dissolving all the reactants can be found [88].

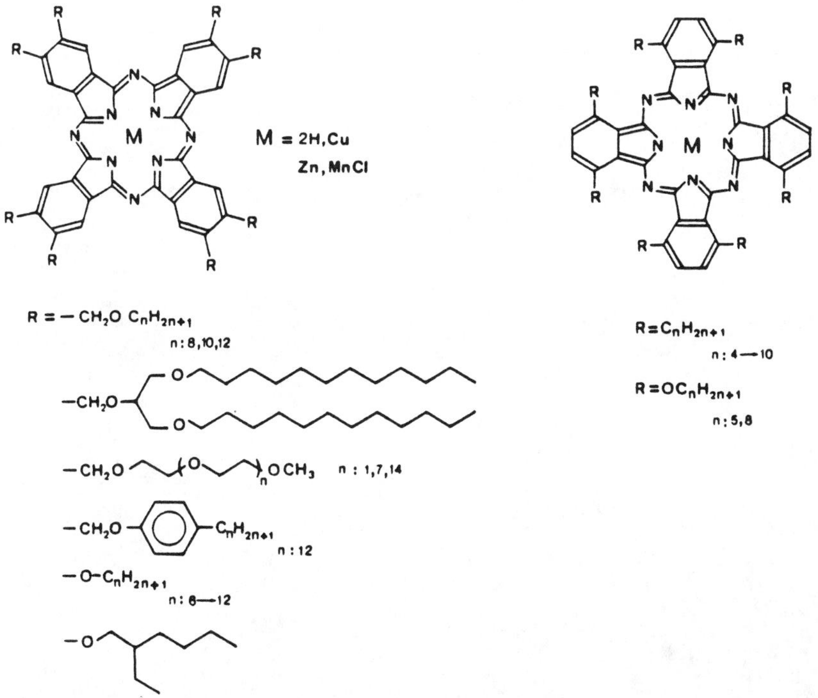

Figure 22 The main phthalocyanine derivatives substituted with the long alkyl chains that have been described (after [12–14, 75, 89]).

Tetrapyrazinoporphyrazine derivatives [170, 171] and octakis (octylthio) tetraazametalloporphyrins [172] have also been shown to form liquid crystalline phases. A cholesteric columnar mesophase has been obtained with optically active side chains [172].

ii. Optical Microscopy and Differential Scanning Calorimetry

Thermotropic liquid crystalline phases are routinely characterized by polarized light microscopy. The texture observed permits one in some cases to determine the type of organization of the mesophase [90]. Columnar liquid crystals are most often highly viscous and become somewhat fluid near the transition temperature to the isotropic liquid. The anisotropic liquid crystal phases show flowerlike textures (Fig. 23) identical to those originally found by Chandrasekhar for substituted benzene derivatives [91].

Additionally, a striated texture was observed for $(C_{12}OCH_2)_8PcH_2$ from 89 to 185°C and $(C_{12})_8PcH_2$ from 80 to 120°C [74]. This transition has been associated with a so-called "Pincement de Skoulios–de Gennes" [29, 30], in which a periodic distortion within the columns is postulated.

Figure 23 Crossed-polarizer optical microscopy photograph of $(C_{12}OCH_2)_8PcH_2$ flowerlike (fan) texture.

This phenomenon is more probably associated with the slow appearance of a highly organized phase [74] or a deviation from hexagonal order due to the tilting of the Pc macrocycle relatively to the column axis [168].

In the case of 3,6-dialkyl-substituted Pc, a fan (flowerlike) texture, a needle texture, and two mosaic textures [14] have been observed.

The phase transitions may also be detected by differential scanning calorimetry. The main contribution to the heat of fusion from crystal to mesophase comes from the melting of the paraffinic chains. The heat of fusion per CH_2 group is strongly dependent upon the number of C atoms for short chains; for chains longer than C_{12}, the increase per CH_2 group is lowered, and the final values are those of polyethylene ($\Delta H = 58.7$ cal/g; mp = 135°C) (Fig. 24).

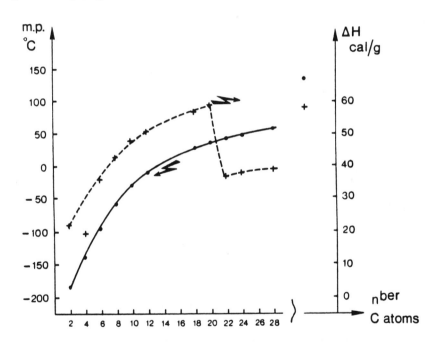

Figure 24 Variation of the heat of fusion (ΔH in cal/g) and of the melting point (in °C) as a function of the number of carbon atoms for paraffins (after[92]). The reference axes are indicated by arrows.

A value of approximately 50 cal/g is expected for C_{10}–C_{18} linear chains [92]; as a comparison, benzene, naphthalene, and anthracene have heats of fusion around 30–35 cal/g (benzene: 30.45; naphthalene: 34.06

cal/g). In $(C_{12}OCH_2)_8PcH_2$, 75% of the mass is constituted of the aliphatic side chains and 25% of the aromatic rigid core [93]; the heat of fusion from the crystal to the mesophase therefore mostly corresponds to the melting of the paraffinic side chains. The calculation for $(C_{12}OCH_2)_8PcH_2$ (M_W: 2101) leads to an upper limit of 20 cal/g for the melting of the paraffinic chains, far below the values expected for linear chains or for smectic liquid crystals [94] but in agreement with the values found in columnar mesophases [91, 95]. This indicates a fairly disordered state of the paraffinic chains within the crystalline phase.`

It is generally observed that the clearing-point temperatures — the passage from the mesophase to the isotropic liquid — decrease with increasing chain length. This is also the case for substituted phthalocyanines (Table 4).

Table 4 Transition Temperatures (°C) and Corresponding Enthalpy Changes (kcal/mole) for the Mesogens R_8PcH_2 [a].

		K_1	K_2	M_1	M_2	I
$R=-CH_2OC_nH_{2n+1}$						
$n=8$	1		67(12)		320	
$n=12$	2		77(27)		260(1)	
$n=18$	3		62(50)		193	
$R=-OC_nH_{2n+1}$						
$n=6$	4		102(12)		>400	
$n=8$	5		94(18)		dec.	
$n=12$	6		91(31)		>300	
$R=-OCH_2-CHC_4H_9$						
$\quad\quad\vert$	7		170(1.1)	223(1.8)	270(0.8)	
$\quad\quad C_2H_5$						
$R=-C_nH_{2n+1}$						
$n=8$	8	124(4)	186(14)		329(2)	
$n=10$	9	137(0.8)	162(12)		281(2)	
$n=12$	10	81(14)	120(<0.5)		252(0.5)	

[a] K: Crystal; M: mesophase; I: isotropic liquid from [74, 76, 88, 89].

The clearing point of $(C_nH_{2n+1}OCH_2)_8PcH_2$ decreases from 320°C (n=8) to 193°C (n=18) with increasing chain length. The same behavior is observed for 3,6-disubstituted Pc [14]: For the shortest chains (n=4,5) no mesophase is found; the M→I transition temperatures decrease linearly with higher chain length. The stability domains of the mesophases also decrease when the number of carbon atoms in the side chains is increased from 253°C (1 ; $n = 8$) to 131°C ($n = 18$).

Branched side chains lead to a considerable decrease of the enthalpy of transitions K→M; however, the structures of the mesophases are different from those of linear chains, making the comparison.difficult. The relative enthalpies of fusion of normal or branched alkanes (*n*-octane: 4.9 kcal/mole; 3-methylheptane: 2.7 kcal/mole; 4-methylheptane: 2.6 kcal/mole), show the same effect.

The formation of metal complexes usually extends the liquid crystal stability domain; for $(C_{12}OCH_2)_8PcM$ the mesophases are stable from 53 to more than 300°C [M=Zn(II)] and from 78 to more than 300°C [M=Cu(II)] [96]. Similar results are found for 3,6-disubstituted complexes [14].

iii. X-Ray Diffraction at Small Angles

Conventional x-ray diffraction at small angles permits one to measure periodicities from 4 to 50 Å, approximately. These determinations are generally sufficient to postulate a plausible structure (Fig. 25).

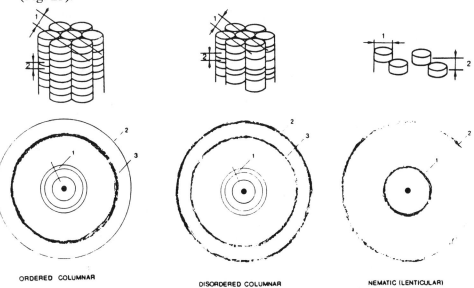

ORDERED COLUMNAR   DISORDERED COLUMNAR   NEMATIC (LENTICULAR)

Figure 25 Characteristic x-ray diffraction patterns corresponding to ordered columnar, disordered columnar, and nematic lenticular mesophases. The assignment of the lines is indicated by numbers. Peak number 3 corresponds to the interparaffinic distance.

The distances in the real space d_r may be obtained from the diameter d of the rings on planar films via the well-known formula :

$$\tan 2\theta = d/2D \tag{18}$$

$$2d_r \sin \theta = \lambda, \qquad \lambda = 1.54\,\text{Å (Cu)}. \tag{19}$$

where D is the sample-to-film distance.

The interparaffinic mean distance is usually associated with a more or less diffuse halo centered around 4.3 Å. The macrocycles may show either a regular spacing or a quasiliquid state within the columns. In the first case, a narrow peak in the region 3.4–3.8 Å is observed.

The intercolumnar distance a may be calculated from the spacing noted for the inner rings. The rays corresponding to a hexagonal lattice are given by :

$$d_r = \frac{a}{\sqrt{4/3(h^2+k^2+hk)}} \tag{20}$$

where a is the lattice parameter and h, k are the Miller indices.

The periodicities observed for the lower Miller indices are indicated in Fig. 26.

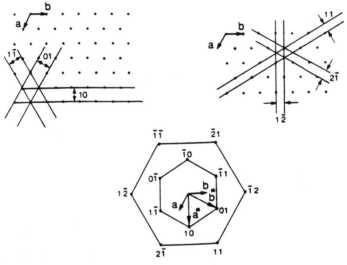

Figure 26 Assignment of the first two inner rays observed for hexagonal lattices. The corresponding x-ray pattern for oriented samples is represented below.

The two-dimensional hexagonal lattice is disrupted when the axis of the column is not normal to the macrocyclic plane. The molecular units may slip relative to each other: a ^2D lattice can be preserved within the plane of the molecular unit, but the axis of the column is no longer normal to the molecular plane, and the lattice becomes rectangular or oblique (Figs. 27 and 28).

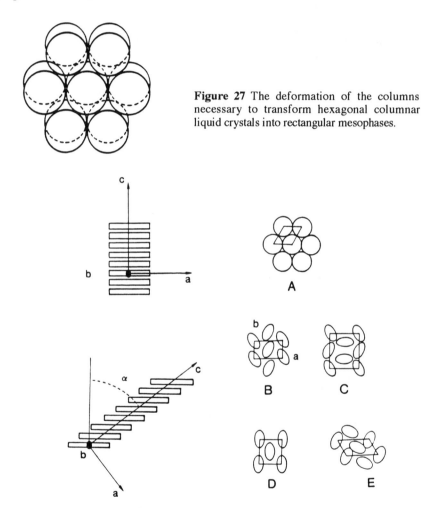

Figure 27 The deformation of the columns necessary to transform hexagonal columnar liquid crystals into rectangular mesophases.

Figure 28 The various types of packing found in columnar mesophases (a) hexagonal; (b)–(d) rectangular ($P2_1/a$, $P2/a$, $C2/m$); (e) oblique (P_1) (after [97, 98]) The figures (b)–(e) represent the projection of the disks belonging to the columns upon the *ab* plane perpendicular to the axis of the column.

The previous features may be found in the x-ray patterns of three differently substituted metal free phthalocyanine derivatives (Table 5).

The polycrystalline solid samples may be indexed by postulating an orthorhombic lattice. A solid-to-solid phase transition modifying the tilting angle has been noticed for $(C_{12})_8PcH_2$ [74]. In the mesophase, the intercolumnar distance for dodecyl side chains varies from 29 to 35 Å, depending upon the connecting link to the phthalocyanine ring. The nature of the atom connecting the side chain to the aromatic core is important for determining the local conformation of the chain [76]. The use of –OR groups leads to larger intercolumnar spacings than –CH$_2$–OR groups (Table 5).

Table 5 Structure and Cell Parameters of R_8PcH_2 (R=–CH$_2$OC$_{12}$: 2 ; R=–OC$_{12}$: 6 ; R=–C$_{12}$: 10) [a]

R=	
–CH$_2$OC$_{12}$H$_{25}$	2
–O–C$_{12}$H$_{25}$	6
–C$_{12}$H$_{25}$	10

	K$_1$	K$_2$	M
2	orthorhombic a=24.6 b=19.4 c=4.3		hexagonal D=31 L=4.6 h=- Λ=20.5
6	orthorhombic a=27.8 b=25.8 c=4.3		hexagonal D=35 L=4.5 h=3.4
10	orthorhombic a=25.1 b=18.8 c=4.3	orthorhombic (100°C) a=23.7 b=20.3 c=4.3	hexagonal D=29 L=4.6 h=-

[a] D: intercolumnar distance; L: correlation length of disordered side chains; h: stacking period along the columns; Λ: see [32, 168].

The influence of the chain length on the intercolumnar distance has been studied (Table 6).

Table 6 Influence of the Chain Length on the Intercolumnar Spacings (Disordered Columnar Mesophases) [a]

$(C_nH_{2n+1}OCH_2)_8PcH_2$	D	L	D_{calc}	D_{free}
$n=8$	26	4.5	40–43	35
$n=12$	31	4.6	45–55	38
$n=18$	36	4.7	50–70	41

[a] D: intercolumnar spacing in Å; L: correlation length of the paraffinic side chains. The diameter of the molecular unit is calculated by assuming a size of 15 Å for the macrocycle. Fully elongated and quasimolten side chain distances are used to calculate the upper and lower limits of D_{calc}, respectively. D_{free} assumes a free rotation state.

The intercolumnar distances found in the mesophases are significantly smaller than the calculated diameter of the molecular units. Consequently an important interlacing of the side chains between contiguous columns or a synclinal intracolumnar arrangement of the paraffin chains must be postulated (Fig. 29).

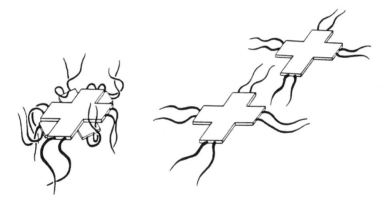

Figure 29 Intercolumnar or intracolumnar interlacing of the paraffinic chains.

Alternatively a tilting of the macrocycle relative to the axis of the column may be assumed [168]. In $(C_{12}O)_8PcH_2$, an ordered hexagonal mesophase is formed in which the intermacrocycle distance is approximately 3.4 Å and the intercolumnar distance 35 Å; for the slightly longer side chains $(C_{12}OCH_2)$, the apparent intercolumnar distance is smaller (31 Å). In the first case, the Pc macrocycle is normal to the column axis, whereas in the second case, a tilting angle could occur [168].

The area (S) occupied by each flexible chain at its anchoring point on the aromatic macrocyle may be calculated. By assuming a cylindrical surface of radius R, it is given by [93] :

$$S = 2\pi Rh/8 \qquad\qquad (21)$$

with $R = 7.5$ Å. S is of the order of 20 Å^2 when h, the intermacrocyclic distance is 3.4 Å. The area per chain varies from 18 Å^2 for crystals to 22 Å^2 for conventional smectic phases. The calculated value is therefore in the expected range.

The formation of metal complexes does not influence the structure of the mesophases (Table 7).

Table 7 Parameters Associated with the Hexagonal Columnar Mesophases of $(C_{12}OCH_2)_8PcH_2$ and of the Corresponding Complexes [a]

	D	h	L
H$_2$	31	—	4.6
Cu	31	3.4	4.7
Zn	31	3.6	4.7
MnCl	31	—	4.6

[a] (same notation as in Table 5). Distances in Å. (An extra ray is observed for M=Zn, 21.6 Å, and M=MnCl, 18.9 Å).

The use of polyoxyethylene side chains permits one to obtain columnar liquid crystals at room temperature [13]. $[CH_3(OCH_2CH_2)_2OCH_2]_8PcH_2$ forms mesophases from <25 to 300°C. A hexagonal packing is observed ($D = 25$ Å, $h = 3.5$ Å, $L = 4.7$ Å). The use of polyoxyethylene side chains is interesting since it permits one to obtain microdomains in which various metallic salts may be dissolved.

G. POLAR PHTHALOCYANINE DISCOGENS

Many liquid crystal visualization devices are based upon the orientation of mesogens in electric fields. This necessitates the use of dipolar mesogens.

The synthesis of in-plane polar Pc discogens has been undertaken [25] (Fig. 30).

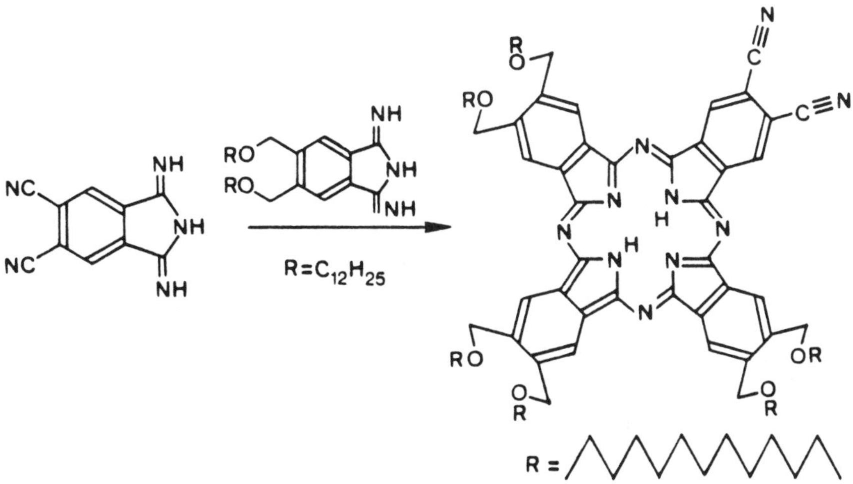

Figure 30 Synthesis of in-plane polar discogens (after [25]).

In-plane polarity necessitates the synthesis of unsymmetrically substituted Pc; only a few reports concern such syntheses [86, 99, 100]. The best pathway uses as an intermediate the monoactivated derivatives of 1,2,4,5-tetracyanobenzene [25]. A 20% yield of isolated unsymmetrical phthalocyanine is thus obtained.

The electrical dipole moment of the molecular unit may be estimated from the vectorial addition of group dipole moments [101]. A value of 7 Debye is found.

The ^1H NMR spectra of the discogen in $(CDCl_2)_2$ at various temperatures (20–90°C) show a strong aggregation tendency of the mesogen. The monomeric form is obtained for dilute solutions; a single peak is obtained for the benzylic protons; at higher concentrations the aggregated species present three peaks. The symmetrically substituted dodecyloxy derivative does not lead to a splitting of the NMR signals even aggregated [13].

By comparison with NMR results obtained for siloxanephthalocyanine dimers or oligomers [102], these observations may be interpreted as a consequence of the slow rotation between adjacent macrocyclic rings. This barrier to rotation must also be present in condensed phases. $(C_{12}OCH_2)_6Pc(CN)_2$ presents a mesophase from –40 to more than 300°C [25, 96]; x-ray diffraction indicates a hexagonal columnar structure with an intercolumnar spacing of 32 Å. A rather sharp peak at 3.4 Å demonstrates a high degree of order of the macrocyclic cores within the columns and a staggered conformation of the macrocycles (Fig. 31).

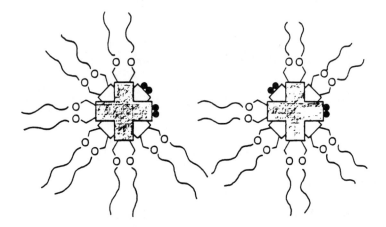

Figure 31 Two of the staggered conformations that probably occur in the columnar mesophases of $(C_{12}OCH_2)_6Pc(CN)_2$ (after [88]).

The overall ^2D hexagonal symmetry of the mesophase implies a random distribution of the dipoles in the columns (Fig. 32).

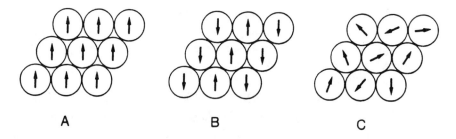

Figure 32 Different ways of organizing polar molecular units in a hexagonal sublattice. (a) Ferroelectric order; (b): antiferroelectric order; (c): random distribution.

Out-of-plane polar molecular units have also been prepared [104]. Divalent lead and tin ions have radii too large (1.2 and 0.93 Å, respectively) to fit the central cavity of the phthalocyanine ring. Out-of-plane complexes are consequently formed in which the metal ion forms a protuberance above the molecular plane of the phthalocyanine (Fig. 33).

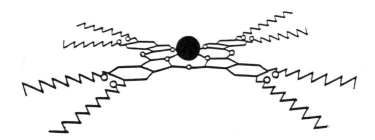

Figure 33 Representation of out-of-plane divalent complexes of $(C_nOCH_2)_8PcM$, M = Sn(II), Pb(II) (after [104]).

The dipole moment of the lead phthalocyanine subunit has not been experimentally determined; dielectric measurements are in progress to measure the molecular dipole moment and the Kirkwood factor of the material [103].

The domain of stability of the mesophases has been determined for $(C_nOCH_2)_8PcPb$ with n = 8, 12, 18 by polarized light microscopy and differential scanning calorimetry (Table 8).

Table 8 Transition Temperatures in °C Determined by DSC for Octasubstituted Phthalocyanine–Lead(II) Derivatives, $(C_nOCH_2)_8PcPb$.

n	K	M	I
8	−45 (0.5)	155	
12	−12 (6.5)	125	
18	46^a $(51)^b$	60^c	

In parentheses are indicated the corresponding enthalpies of transition (kcal/mole).
[a]The two peaks may overlap depending on heating rate.
[b]This value represents the sum of the K →M and M→I enthalpies of transition.
[c]K: crystal; M: mesophase; I: isotropic liquid.

The mesophase-to-isotropic liquid transition temperatures (M→I) decrease with increasing chain length. The crystal-to-mesophase transitions are situated below room temperature for the two shortest chains. The corresponding enthalpies are, however, surprisingly low. The textures observed by optical microscopy are characteristic of columnar mesophases. The octadecyl derivative leads to well defined textures only by cooling from the isotropic liquid; the liquid crystalline phase is metastable even in the range 46–60°C.

X-ray diffraction demonstrates the columnar structure of the mesophases (Table 9).

Table 9 Geometrical Characteristics of the Hexagonal Columnar Mesophases of $(C_nOCH_2)_8PcPb$[a]

n	D	h
8	26.9	7.4
12	31	—
18	36	4.5(?)

[a] D: intercolumnar distance; h stacking period along the columns (after [104, 105]).

An important point concerns the periodicity within the columns. For the octyl derivative, a broad band is observed around 7.4 Å. This band is probably due to the presence of pairs of molecules as encountered in the triclinic structure of unsubstituted PcPb [106–108].

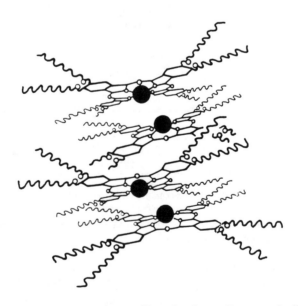

Figure 34 Schematic representation of the antiferroelectric coupling occurring in the liquid crystalline phase of $(C_8OCH_2)_8PcPb$ (after [104, 105]).

In the case of $(C_{18}OCH_2)_8PcPb$ a halo centered around 4.5 Å is observed; no peak can be found in this region for $(C_{12}OCH_2)_8PcPb$. The type of one-dimensional ordering therefore seems to depend on the chain length of the side groups. A rough estimate of the periodicity parameter within the columns can be made knowing the van der Waals and ionic radii of the components (Fig. 35).

Taking into account the protuberance due to the lead ion, the thickness of the molecular unit is in the range 3.8–4.2 Å. An untilted ferroelectric ordering [Fig. 35(a)] should lead to a periodicity of ≈4 Å, whereas the antiferroelectric homologue would give the double value [Fig. 35(c)]. The tilted conformations [Fig. 35(b) and (d)] lead to spacings of $4/\cos \alpha$ and $8/\cos \alpha$, respectively.

The value observed at 7.4 Å is compatible with an antiferroelectric arrangement within the columns. In $(C_{18}OCH_2)_8PcPb$, the halo at 4.5 Å can be due either to the intermacrocycle distance in a ferroelectric arrangement or, more probably, to the correlation length of the disordered aliphatic chains.

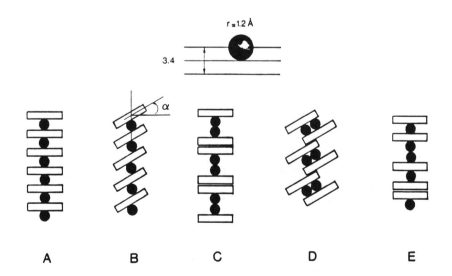

Figure 35 Estimation of the geometrical parameters of PcPb molecular subunits and some of the possible packings.

The lead complex is fairly unstable, and ion exchange readily occurs in weakly acidic media [104]. Hamann and co-workers [109, 110] have reported that in the monoclinic modification of PcPb, where the macrocycles are untilted, lead(II) ion exchange may be promoted by electric fields to form highly conducting materials. The nature of this phenomenon must, however, be more accurately determined since thermal effects cannot be excluded. The same type of ion exchange between the macrocycles may be envisaged with the previous mesogens.

The PcPb subunit can also lead to new types of chiral molecular units. The symmetry of PcPb is C_{4v}. It is transformed into a chiral group by breaking the symmetry within the macrocyclic plane (Fig. 36).

Tin(II) yields complexes of similar geometry. However, the stannous derivative is not stable in air and is readily transformed into the stannic compound at room temperature in solution [104]. This transformation may be followed by optical absorption spectra. The unoxidized species absorbs at 720 nm (blue-violet color) at the same wavelength as the lead derivatives. In presence of air, a chloroformic solution of the Sn(II) complex is quantitatively transformed to a new species absorbing around 700nm ; this last value is characteristic of $PcSn(IV)X_2$ derivatives [111] (Fig. 37).

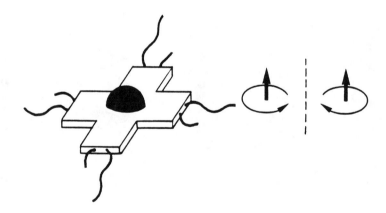

Figure 36 Chiral out-of-plane phthalocyanine complexes.

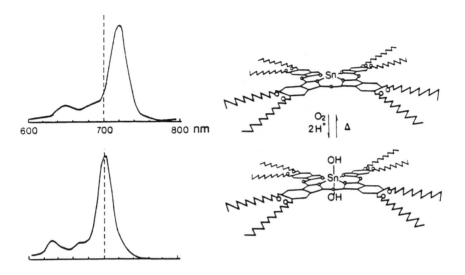

Figure 37 Optical absorption spectra of $(C_{12}OCH_2)_8PcSn$ and of the corresponding stannic derivative (after [104]).

The dihydroxo-derivative is reduced back to the tin (II) complex by heating at 130°C. The dihydroxo-derivative may be used for obtaining polyoxystannylpolymers [112] in which a covalent backbone (spine) is surrounded by alkyl-substituted macrocycles (spinal columnar liquid crystal see Fig. 38).

H. SPINAL COLUMNAR LIQUID CRYSTALS

It is well known that oxystannyl [113,114] and oxysilyl polymers [113–117] may be formed by heating in vacuo the corresponding dihydroxymetal (IV) phthalocyanines. The same type of polycondensation takes place in columnar liquid crystalline phases (Fig. 38).

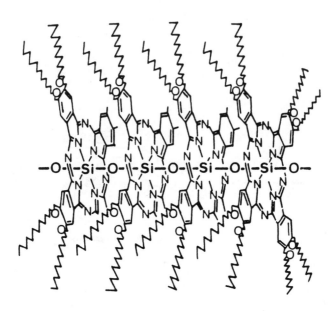

Figure 38 Spinal columnar liquid crystals formed by condensation of $(C_{12}OCH_2)_8PcSi(OH)_2$.

The dihydroxo-tin(IV) phthalocyanine derivative was obtained in two steps [112] starting from $(C_{12}OCH_2)_8PcH_2$. This latter was treated with $SnCl_2$ in refluxing pentanol to give quantitatively $(C_{12}OCH_2)_8PcSn$, which, when reacted with H_2O_2, yielded $(C_{12}OCH_2)_8PcSn(OH)_2 \cdot 2H_2O$. The liquid crystalline properties of this compound were studied by conventional techniques (DSC, optical microscopy, x-ray diffraction at small angles) [112]. A transition is observed at 59°C from a highly ordered crystalline phase to a liquid crystal (Table 10).

Table 10 Mesomorphic Properties of $(C_{12}OCH_2)_8PcSn(OH)_2 \cdot 2H_2O$ [a]

K	M$_1$	M$_2$	I
59 (8.1)	95 (1.8)	114 (9.7)	
orthorhombic	rectangular		
$a=45.4$	$a=25.2$		
$b=30.6$	$b=30.7$		
$c=3.8$ (?)	halos:4.4		
	3.8		

[a]Temperatures in °C, in parenthesis the enthalpies of transition (kcal/mole).

At 95°C a new texture slowly appears with a probable concomitant loss of water. The isotropic liquid is formed at 114°C. The solid phase, at room temperature, presents an x-ray pattern compatible with an orthorhombic structure; it transforms into a rectangular mesophase at 59°C. Simultaneously, the two intense and narrow peaks centered at 4.4 and 3.85 Å are converted into diffuse halos. The first signal is probably related to the interalkane chain spacing, whereas the second one may be associated with the intermetallic distance within the columns.

The orthorhombic lattice (two units per cell) is transformed into a rectangular lattice (one unit per cell) with an increase of volume of approximately 10% (Fig. 39). In the second case, the columns are all tilted in the same direction. The transformation between the two lattices does not involve such a large displacement of the axes of the columns as expected from the very different lattice parameters.

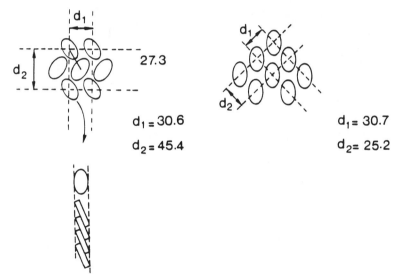

Figure 39 Two of the lattices that can be associated with $(C_{12}OCH_2)_8PcSn(OH)_2 \cdot 2H_2O$.

The fluid isotropic liquid slowly transforms into a highly viscous partially anisotropic mass with a change of color from green to yellow green. This was interpreted as being due to the formation of oxystannyl polymers. ^{1}H NMR, UV–visible absorption spectra, and IR spectroscopy substantiated this assignment [112].

A very similar behavior is obtained for silicon derivatives. The dihydroxysilicon(IV) phthalocyanine derivative $(C_{12}OCH_2)_8PcSi(OH)_2$

is obtained by reacting the corresponding 1,3-diimino-isoindoline with silicium tetrachloride in quinoline at 150°C; the dichlorosilicon(IV) derivative is hydrolyzed with water at room temperature to give the desired molecule [118].

Optical microscopy shows a birefringent viscous mass from room temperature to 300°C with no visible transition. DSC reveals a transition at −7°C (15 kcal/mole). $(C_{12}OCH_2)_8PcSi(OH)_2$ is therefore liquid crystalline at room temperature. X-ray diffraction measurements at room temperature and at 150°C both show a series of five Bragg reflections in accordance with a hexagonal lattice (intercolumnar spacing 30.8 Å). An extra line is seen at 19.8 Å.

Polycondensation is achieved by heating $(C_{12}OCH_2)_8PcSi(OH)_2$ in the liquid crystalline phase at 180°C for 7 hours. Polymerization was followed by optical absorption spectroscopy and gel permeation chromatography.

The electronic spectra of the oligomers previously obtained with unsubstituted $PcSi(OH)_2$ are quite sensitive to the number of macrocycles within the polycondensate. Blue shifts are observed for the Q and B bands when the polysiloxane chain length is increased. In the monomer, Q and B are centered at 665 and 351 nm; in the dimer the bands are shifted to 630 and 329 nm; and in the trimer to 618 and 327 nm, respectively [119]. The same behavior is observed in the case of the oligomerization of $(C_{12}OCH_2)_8PcSi(OH)_2$ (Fig. 40).

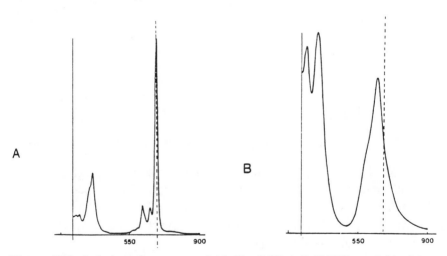

Figure 40 Optical absorption spectra of (a) $(C_{12}OCH_2)_8PcSi(OH)_2$ and (b) of the condensation product obtained by heating it at 180°C for 7 h.

In the samples, heated at 180°C, the peak at 682 nm is replaced by a band around 646 nm. The samples heated for 7 h contained approximately 30% of monomer, 30% of dimer, 20% of trimer, and 20% of higher oligomers, as shown by GPC experiments (Fig. 41).

X-ray patterns of the heated samples are significantly modified as compared to the monomer. A series of three Bragg reflections with reciprocal spacings in the ratio 1:2:3 is indicative of a lamellar order (interlamellar distance, 31 Å). This is not modified from RT to 60°C. Two broad and diffuse outer rings corresponding to 4.5 and 3.4 Å are observed. For temperatures higher than 60°C, optical microscopy and x-ray diffraction both show the formation of an isotropic liquid [118].

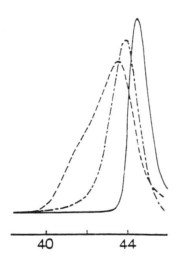

40   44

Figure 41 Gel permeation chromatograms of, from right to left: $(C_{12}OCH_2)_8PcSi(OH)_2$, the same compound heated for 1 h at 180°C, and 7 h at 180°C.

The transformation of the hexagonal lattice into a lamellar one that accompanies the polycondensation process does not necessarily involve important organization changes within the mesophase (Fig. 42). A slight deformation of the array of columns is sufficient to break the hexagonal symmetry while preserving most of the characteristic distances of the hexagonal lattice.

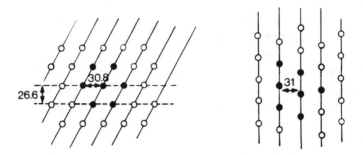

Figure 42 Transformation of a hexagonal lattice (intercolumnar distance, 30.8 Å) into a lamellar array (interlamellar distance, 31 Å).

I. ENERGY TRANSPORT IN COLUMNAR LIQUID CRYSTALS

The optical absorption characteristics of phthalocyanine and metallophthalocyanine subunits are well documented [33–36] see also [173]. Two main bands centered around 700 nm (Q band) and 350 nm (B band) correspond to $\pi \to \pi^*$ transitions. In the case of PcH$_2$, the symmetry is lowered from D_{4h} to D_{2h}, leading to a splitting of the Q band.

At low temperatures, the two inner hydrogen atoms cannot exchange over long periods of time, and the transfer may be photochemically induced. This phenomenon has been used for making photochemical hole-burning devices [120–122].

By studying the luminescence properties of solutions of mixtures of $(C_{12}OCH_2)_8PcH_2$ and $(C_{12}OCH_2)_8PcCu$ [124], it has been demonstrated that the copper complex quenches the singlet excited state of the metal-free derivative. The maximum of the fluorescence intensity was also shown to be displaced to long wavelengths in aggregates or condensed phases [124]. Similar red shifts were reported in the case of the aggregation of porphyrin derivatives [126, 127]. The energies corresponding to excited states of unsubstituted PcH$_2$ and PcCu can be found in the literature [123, 125].

The distribution of the copper complex within solid phases of $(C_{12}OCH_2)_8PcH_2$ was studied by electron paramagnetic resonance [128].

The EPR pattern of $(C_{12}OCH_2)_8PcCu$ involves the coupling of the unpaired electron of Cu(II) with the nuclear spin of copper ($I=3/2$) and with the four equivalent nitrogen atoms ($I=1$). When the copper complex is diluted in a diamagnetic matrix, the expected pattern with 4×9 lines is obtained.

EPR spectra of the pure copper compound consist of one broad asymmetrical line due to dipolar and exchange interactions between the paramagnetic copper ions [128] (Fig. 43). In this case a signal is observed at half-field corresponding to $\Delta M = \pm 2$.

The spectrum of the copper complex is therefore significantly different when isolated or when it interacts with another copper complex. When the two species coexist, due to the difference in linewidths, only the spectrum of the isolated species can be observed; this permits one to determine the distribution of the copper complex incorporated within columns of the metal-free derivative.

In the one-dimensional case, the concentration of A isolated between two B molecular units is given by [129, 130] :

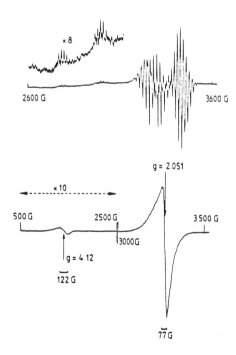

Figure 43 EPR spectra of $(C_{12}OCH_2)_8PcCu$ diluted in $(C_{12}OCH_2)_8PcH_2$ (1.7% m/m) and pure $(C_{12}OCH_2)_8PcCu$ (dimers, aggregates, or undiluted).

$$[BAB] = x_A[1+(x_A/x_B)]^{-2} \qquad (22)$$

where x_A and x_B are molar fractions of A and B, respectively.

For a concentration of 18% of A in B, only 12% appear as isolated species, the remaining 6% being essentially present as dimers or aggregates. By varying the concentration of $(C_{12}OCH_2)_8PcCu$ into the metal-free derivative, it is possible to determine that the amount of isolated copper complex approximately follows Eq. (22) and that, in consequence, a random distribution can be assumed.

The luminescence properties of mixtures of $(C_{12}OCH_2)_8PcCu$ in the metal-free derivative have been studied at different molar ratios [130] (Fig. 44). Concentrations of Cu(II) complexes not inferior to 0.05% (m/m) could be attained. The metal-free derivative is always contaminated with copper complexes because CuCN is used to transform the dibromo derivative into the corresponding phthalonitrile during the synthesis.

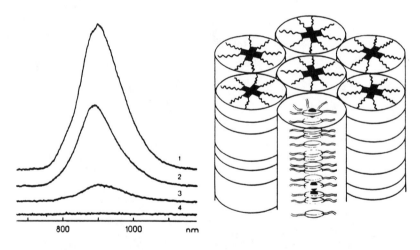

Figure 44 Fluorescence of mixtures of $(C_{12}OCH_2)_8PcCu$ into $(C_{12}OCH_2)_8PcH_2$ at room temperature (excitation wavelength, 514.5 nm); 1:0.05% (m/m); 2:0.4%; 3:4.5%; 4:16%.

For only 16% of copper complex, the luminescence of $(C_{12}OCH_2)_8PcH_2$ is almost entirely quenched. Energy migration processes must therefore intervene in the quenching of fluorescence (Fig. 45).

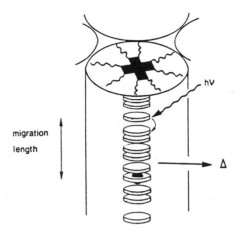

Figure 45 Schematic representation of the absorption of photons in columnar mesophases. Quenching occurs in the vicinity of the copper complex.

A model was established to calculate the exciton diffusion length within the columns. The probability $N(n)$ of having a sequence of n metal-free molecules (A) between two copper complexes (B) is given by :

$$N(n) = P_{ba}(P_{aa})^{n-1}P_{ab} \qquad (23)$$

where P_{ba} is the probability to have species A after species B.

The various probabilities may be calculated knowing the molar fractions x_A and x_B :

$$P_{aa} = P_{ba} = x_A \qquad (24)$$

$$P_{ab} = x_B = 1 - x_A \qquad (25)$$

and then

$$N(n) = (x_A)^n(1 - x_A) \qquad (26)$$

The probability that a photon is absorbed by a molecule A belonging to a sequence of length n is given by :

$$\frac{nN(n)}{\displaystyle\sum_{n=1}^{\infty} nN(n)} \qquad (27)$$

The intensity of fluorescence in the presence of quencher I_q is proportional to the number of photons absorbed multiplied by $(n-n_c)/n$ if $n > n_c$ and 0 if $n \leq n_c$. n_c is the exciton diffusion length (Fig. 46).

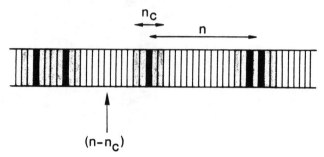

Figure 46 Model used for the luminescence properties of one-dimensional systems containing a randomly distributed quencher [130]. • quencher, * chromophore, n: number of molecules between two quenchers; n_c : extent of energy migration.

$$I_q = I_0 \, \frac{nN(n)}{\displaystyle\sum_{n=1}^{\infty} nN(n)} \, \frac{n-n_c}{n} \qquad n > n_c \qquad (28)$$
$$ 0 \qquad\qquad\qquad n \leq n_c$$

The amount of quenching $(I_q/I_0)_{tot}$ is the sum of the previous equations over all the sequences:

$$\left(\frac{I_q}{I_0}\right)_{tot} = \sum_{n=1}^{\infty} \frac{nN(n)}{\displaystyle\sum_{n=1}^{\infty} nN(n)} \, \frac{n-n_c}{n}, \qquad n > n_c \qquad (29)$$
$$\phantom{\left(\frac{I_q}{I_0}\right)_{tot} =} 0 \qquad\qquad n \leq n_c$$

$$\left(\frac{I_q}{I_0}\right)_{tot} = \frac{\displaystyle\sum_{n_c+1}^{\infty} (n-n_c)N(n)}{\displaystyle\sum_{n=1}^{\infty} nN(n)} \qquad (30)$$

It can then be readily demonstrated that:

$$\left(\frac{I_q}{I_o}\right)_{tot} = (x_A)^{n_c} \qquad (31)$$

Experimental points have been fitted to Eq. (31), leading to an exciton diffusion length of 100–500 Å. In unsubstituted phthalocyanine solid thin films, singlet exciton migrates over a distance of 300–500 Å [131].

The luminescence of $(C_{12}OCH_2)_8PcH_2$ has also been studied at a temperature higher than 80°C, which corresponds to the appearance of the mesophase [130]. A strong fluorescence decrease is observed when going from the crystal to the liquid crystalline phase. The low fluorescence intensity can be related either to a more disorganized state in the mesophase, leading to the creation of traps, or to a more extended migration energy path.

Laser excitation of $(C_{18}OCH_2)_8PcH_2$ in the solid and liquid crystalline phases [132] gives transient differential absorption spectra typical of a phthalocyanine triplet–triplet absorption [123]. The lifetime of the excited species in benzene solutions ($10^{-6} M$) is 120 ± 10 μs in agreement with the values reported for the nonsubstituted derivative [123, 133]. For low-energy laser pulses in the solid state, a monoexponential decay of the transient species is observed, with a corresponding triplet lifetime of 7.5 ± 0.5 μs [132] at 25°C. This lifetime decreases to 4.2 ± 0.3 μs in the mesophase (84°C). For high-energy pulses, the transient decay obeys a second-order kinetic law, demonstrating the efficiency of a triplet–triplet annihilation process : $T+T \rightarrow 2S_0$. In this case also, energy migration occurs within the columns of the discotic liquid crystal. The exciton migration length has been experimentally determined, ignoring [132] or taking into account [134] the dimensionality of the diffusion process.

J. CHARGE TRANSPORT IN COLUMNAR MESOPHASES

The intrinsic generation of charge carriers in a molecular material may be represented by the reaction:

$$AAA \leftrightarrow A^+A^-A \qquad \text{free carriers} \qquad (32)$$

where A is a molecular unit.

The ion pair A^+A^- is thermally or photochemically generated, and the activation energy for its formation may be estimated from the redox potentials of the molecular units in solution [135] :

$$A \leftrightarrow A^+ + e^- \quad E_{ox} \tag{33}$$

$$A + e^- \leftrightarrow A^- \quad E_{red} \tag{34}$$

$$E_{act} = e(E_{ox} - E_{red}) + E_p \tag{35}$$

E_p corresponds to the difference in solvation energy of the reacting species between the solution and the solid or liquid crystalline phases. This equation is valid whenever the interaction energy between the molecular units is small enough as to neglect collective electronic states. The difference $\Delta E = e(E_{red} - E_{ox})$ is of the order of 2 eV for most divalent complexes of phthalocyanines [67]. The corresponding concentration of charge carriers may be estimated from the relation :

$$n = n_0 \ \exp(-\Delta E/2kt) \tag{36}$$

$n_0 \approx 10^{21}/cm^3$ is the concentration of molecular units in the solid state and n is the concentration of charge carriers, A^+ or A^-.

With $\Delta E = +2$ eV, Eq. (36) leads to a concentration of charge carriers of 1.5×10^4 per cm^3, 10^{-11} ppm (m/m). The conductivity is related to the concentration of charge carriers and to the mobility by :

$$\sigma = ne\mu \tag{37}$$

where $e = 1.6 \times 10^{-19}$ Coulomb.

The mobility of charge carriers varies from 10^{-5} to 1 $cm^2/V.s$, depending on the degree of disorder in the material; the conductivity is therefore between 10^{-15} and 10^{-20} $\Omega^{-1} cm^{-1}$ when $\Delta E = 2$ eV.

It is possible to obtain a higher concentration of charges by introducing electron acceptors or electron donors. In this case, doped insulators with semiconducting properties are obtained [67]. All low-molecular-weight compounds and polymeric systems studied so far behave as insulators or doped insulators [67, 136, 137]. Phthalocyanine radicals are the only exceptions [137–140]. Bis(phthalocyaninato)lutetium (Pc$_2$Lu) and lithium monophthalocyanine (PcLi) can be both easily oxidized and reduced with $\Delta E = 0.48$ and 0.83 eV, respectively. The intrinsic conductivities of thin films or single crystals of radical phthalocyanines are consequently in the semiconducting range (10^{-6}–10^{-1} $\Omega^{-1}\,cm^{-1}$).

Pc$_2$Lu is an intrinsic molecular semiconductor with a small interaction energy between the molecular units (35 meV), and it is relatively insensitive to structural disorder [137]. PcLi is on the contrary a large-band semiconductor with strong interunit overlap (>1 eV); it is highly affected by structural changes [136–140]. The discogens based on the corresponding molecular subunits have been prepared.

i. Lutetium and Lithium Phthalocyanine Discogens

Unsubstituted lutetium bisphthalocyanine was first synthesized by Kirin and co-workers [141] by heating a mixture of phthalonitrile and lutetium acetate. The corresponding octasubstituted derivatives are obtained either by reacting the lutetium(III) salt with the dianion s-Pc^{2-} [142] or by the reaction of the substituted phthalonitrile with Lu(OAc)$_3$ in the presence of 1,8-diazabicyclo [5.4.0] undec-7-ene in n-hexanol [143] (Fig. 47).

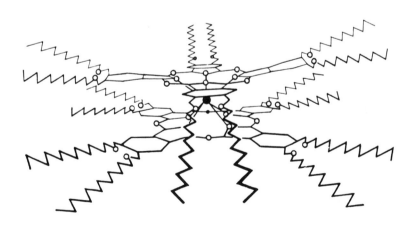

Figure 47 Molecular structure of lutetium bisphthalocyanine substituted with dodecyloxymethyl side chains.

EPR studies have shown that one of the macrocycles is a radical and that the formula can be written $Pc^{2-}Pc^{-\bullet}Lu(III)$ [144–146]; similar results were found for the corresponding discogens [142].

Radicals can be both easily oxidized and reduced: The exceptional electrochemical properties of the Pc_2Lu and $PcLi$ subunits, therefore, derive from their peculiar electronic properties.

The lutetium and lithium discogens form liquid crystalline phases (Table 11).

The domain of stability of the mesophase for $[(C_nOCH_2)_8Pc]_2Lu$ is very narrow or absent depending on the chain length: only the octadecyl derivative gives transitions that can be detected by DSC. When the side chain is linked via an oxygen atom to the macrocycle as in $[(C_nO)_8Pc]_2Lu$, the domain of stability of the mesophase is considerably extended. The textures observed for the lithium and lutetium derivatives were very similar to those previously found. Further characterizations were carried out by x-ray diffraction (Table 12).

Table 11 Mesomorphic Properties of Lutetium Bisphthalocyanine and Lithium Phthalocyanine Discogens

	K_1	K_2	M	I
$[(C_nOCH_2)_8Pc]_2Lu$				
$n=8$			$(25)^a$	
$n=12$		24^b	30^b	
$n=18$		51	56	
$[(C_nO)_8Pc]_2Lu$				
$n=12$	41 (16)	85 (14)	189 (2)	
$(C_nO)_8PcLi$				
$n=12$		92 (20)	269 (dec)	
$(C_nOCH_2)_8PcH_2$				
$n=12$		79 (27)	260 (1)	
$(C_nO)_8PcH_2$				
$n=12$		91 (31)	>300	

[a]Direct transition from K to I.
[b]Transition observed only after a complete cycle of heating and cooling. In parentheses is indicated the enthalpy of the transition (kcal/mole).

Table 12 Structural Parameters Determined by X-ray Diffraction for the Lutetium and Lithium Phthalocyanine Derivatives

	K_1	K_2
$[(C_{18}OCH_2)_8]PcLu$		hexagonal $D=37$ $h=7.3$
$[(C_{12}O)_8Pc]_2Lu$	orthorhombic $a=29$ $b=24.6$	hexagonal $D=34.6$ $L=4.6$ $h=3.3$
$(C_{12}O)_8PcLi$	quadratic $a=b=25.5$	hexagonal $D=34.9$ $L=4.6$ $h=3.35$

Both the dodecyl- and octadecyl-lutetium derivatives form hexagonal columnar liquid crystals. The intercolumnar distances are comparable to those found for the corresponding metal-free derivatives. In the case of $[(C_{18}OCH_2)_8Pc]_2Lu$, a halo is observed at 7.3 Å that can be assigned to the intracolumnar lutetium–lutetium distance. Two narrow lines at 4.3 and 4.1 Å are additionally observed instead of the diffuse band centered at 4.5 Å usually found [142]. These values could arise from partial crystallization of the paraffinic chains.

Surprisingly, an intense and narrow line around 3.3 Å is observed for the mesophase of $[(C_{12}O)_8Pc]_2Lu$ (Table 12). A model involving the insertion of single macrocycles — most probably the corresponding metal-free derivative — within the columns has been suggested by J. Prost (Fig. 48).

The ion–ion periodicity within the columns is disrupted by the metal-free phthalocyanine; the incorporation of the single macrocycle induces a half-period shift, and the periodicity is no longer related to the ion–ion distance. The amount of single macrocycle that must be incorporated to see this effect is related to the coherence length of the x-rays. A value in the range 100–500 Å is expected with the experimental setup used; the metal-free phthalocyanine concentration is therefore at least of the order of 0.5–1%. On the contrary, the interplanar periodicity is not altered by the incorporation of the metal-free derivative, and the corresponding distance will be observed in the x-ray diffraction pattern. The corresponding value (3.30 Å) and the narrowness of the line are, however, quite unexpected for columnar mesophases.

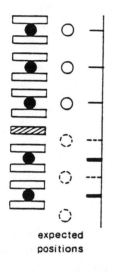

expected
positions

Figure 48 Columnar stacking of bisphthalocyaninato-lutetium. The insertion of a single macrocyclic unit within the columns disrupts the ion–ion periodicity while the interplanar parameter is conserved.

ii. Redox Properties and Electrochromism of the Liquid Crystalline Phases of Lutetium Phthalocyanine Derivatives

The Pc_2Lu unit is known to undergo reversible redox processes associated with color changes: blue for the reduced species Pc_2Lu^-, and orange-red for the oxidized derivative Pc_2Lu^+ [147–152]. The redox potentials in solution of $[(C_nOCH_2)_8Pc]_2Lu$ (n=8,12) have been shown to be very close to those of the unsubstituted complex (Table 13) [153].

Table 13 Formal Potentials (in V versus ferrocene) and Spectra λ_{max} in nm,(ε) in 1 mole^{-1} cm^{-1} of Solution of Lutetium Diphthalocyanines in CH_2Cl_2 [a]

	E ored	E oox	Blue	Green		Orange-Red
$[(C_8OCH_2)_8Pc]_2Lu$	-0.41	-0.04	627 (1.06×10^5)	668 (1.2×10^5)	488 (4.95×10^4)	701 (4.45×10^4)
$[(C_{12}OCH_2)_8Pc]_2Lu$	-0.42	+0.01	631 (1.06×10^5)	671 (1.1×10^5)	492 (4.95×10^4)	704 (4.45×10^4)
Pc_2Lu (151)	-0.45	+0.03	620 (1.55×10^5)	660 (1.55×10^5)	472 (4.7×10^4)	690 (4.7×10^4)

[a] After [153].

The oxidized and reduced species are formed via a fully reversible electrochemical process. They are highly stable and remain unchanged without fading when an external voltage is maintained [153].

All colors may be obtained by the proper combination of the three basic colors: cyan, magenta ,and yellow, or their complementaries to white light: red, green, and blue. In this last case, the white color is obtained by adding the three basic colors (Fig. 49).

The comparison of the theoretical spectra of the three basic colors with the oxidized, neutral, and reduced forms of Pc_2Lu is shown Fig. 50.

The green and blue species of s-Pc_2Lu conform fairly well to the theoretical spectra. On the contrary, the red color is obscured by an additional peak around 700 nm.

Further studies have been made on thin films of $[(C_nOCH_2)_8Pc]_2Lu$ [155]. The redox properties were studied by cyclic voltametry in aqueous solutions in the presence of various salts. The oxidization cycle always leads to well-defined voltamograms, whereas the reduction step is only detected by maintaining the potential during several minutes. Long-term stability under cycling was observed: In the case of $[(C_nOCH_2)_8Pc]_2Lu$, 10^6 cycles led to only a 5% loss of the electrochemical activity [155].

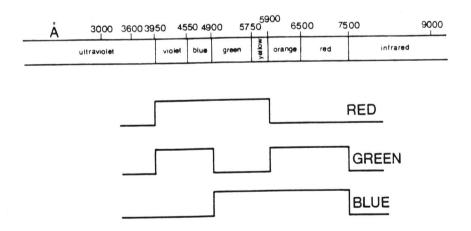

Figure 49 Schematic representation of the absorption spectra of dyes susceptible to give the three basic colors: red, green, and blue (after [154]).

The mesophases exhibited by the oxidized lutetium complexes are more stable than the neutral species (Table 14).

Table 14 Transition Temperatures (in °C) of $[(C_nOCH_2)_8Pc]_2Lu^+$, $SbCl_6^-$, and $[(C_nO)_8Pc]_2Lu^+$, BF_4^-

	K	M	I
$[(C_nOCH_2)_8Pc]_2Lu^+SbCl_6^-$			
n= 8	−10	130	
12	13	118	
18	56	132	
$[(C_nO)_8Pc]_2Lu^+$, BF_4^-			
n= 12	−3(7)	>200	

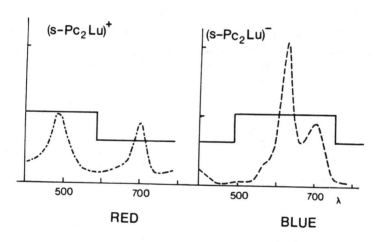

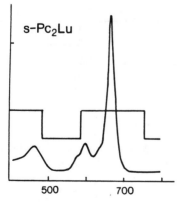

Figure 50 Optical absorption spectra of $[(C_{12}OCH_2)_8Pc]_2Lu$ and of the corresponding cation and anion. The full line represents the theoretical spectra for the green, red, and blue colors.

Neutral $[(C_nOCH_2)_8Pc]_2Lu$ shows a liquid crystalline phase only for the octadecyl derivative and over a very narrow temperature range ($5°C$). All the oxidized forms studied demonstrate a mesomorphic behavior over an extended temperature domain. The mesophase to isotropic transition is almost unaffected by the length of the alkyl side chains, whereas the extent of the mesomorphic domain decreases with increasing chain length. X-ray determinations demonstrated that the mesophases have a hexagonal symmetry with an intercolumnar distance very comparable to those found for the neutral s-Pc_2Lu or the metal-free derivatives (Table 15).

Table 15 X-Ray Characterizations of the Oxidized Forms of the Alkyl-Substituted Lutetium Complexes

	M		
$[(C_{12}OCH_2)_8Pc]_2Lu^+, SbCl_6^-$	hexagonal		
	$D=33.5$	$h=6.75$	
$[(C_{12}O)_8Pc]_2Lu^+, BF_4^-$	hexagonal		
	$D=34.6$	$h=3.27$	$L=4.5$

The columnar structure allows the incorporation of fairly big anions without disrupting the molecular organization. It seems evident that anion–anion and cation–cation electrostatic repulsions should overcome the stabilization arising from cation–anion interactions (Fig. 52). An overlap between partially occupied orbitals of the macrocycles within the columns could stabilize such mesophases. The anions could also induce a high dipole moment of the cationic Pc subunits stabilizing intracolumnar interactions [as suggested by Mr Durand]. Both hypotheses are in agreement with the unusually low values of the intracolumnar periodicities found for $[(C_{12}OCH_2)_8Pc]_2Lu^+$, $SbCl_6^-$ ($h=6.75$ Å), and $[(C_{12}O)_8Pc]_2Lu^+, BF_4^-$ ($h=3.27$ Å $\times$ 2=6.54 Å), smaller than the expected van der Waals distance (6.8 Å).

iii. Electrical Properties

dc and ac electrical determinations over a large frequency range (10^{-3} Hz – 1 GHz) have been carried out with the lutetium and lithium complex discogens.

In the case the system is modeled with two RC circuits in series, when $R_1 >> R_2$, the resistance R_1 is measured at low frequencies while, at high frequencies, both R_1 and R_2 intervene. The same type of results is found for the capacitance terms (Fig. 51).

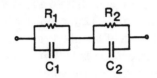

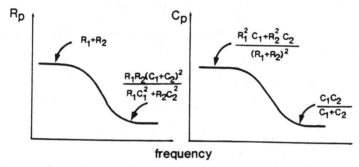

frequency

Figure 51 Resistance (Rp) and capacitance (Cp) as a function of the frequency for the simplest equivalent electrical circuit.

The chemical mechanisms involved in the charge transport must then be associated with each RC circuit. The ac electrical determinations allow one to know the magnitude and the time response constant associated with the chemical processes. In molecular materials, the overall electrical response depends on the electrical properties of various domains (Fig. 52).

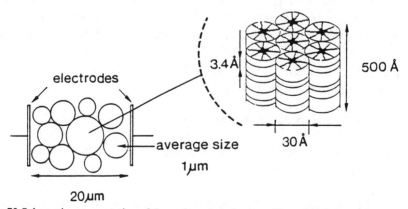

Figure 52 Schematic representation of the various domains found in molecular materials.

The distance between the two electrodes is of the order of 20 μm. Over such a distance, the molecular material is constituted of polydomains or microcrystals whose average size is of the order of 1 μm. The electrical properties will arise from the charge transport within the domains and between the domains. Within a single domain, intracolumnar and intercolumnar charge-transfer processes must be distinguished.

In molecular materials with interunit distances of the order of the van der Waals value (≈ 3.4 Å), the individual units constitute spatially localized electronic states for charge carriers. The following discussion will therefore focus on hopping processes. A considerable number of experimental data concerning the frequency-dependent conductivity of a large variety of materials have shown that most data verify the empirical relationship [157] :

$$\sigma(\omega) = A(T)\omega^{n(T)} \qquad (38)$$

where the index n is between 0.5 and 1 and T is the temperature.

The electrical properties of R_8PcH_2 (R=–OC_{12}; –OCH_2C_{18}), $[R_8Pc]_2Lu$ (R=–OC_{12}; –OCH_2C_{18}), and R_8PcLi (R=–OC_{12}) have been determined [143].

At low frequency (<100 Hz), all the compounds are insulators ($\sigma <$ 10^{-11} Ω^{-1} cm^{-1}. A ω^0 dependence is observed; it is associated with a macroscopic phenomenon where carriers must hop between crystallites of different sizes. In this case, the rate-limiting step for charge carrier hopping is probably related to intercrystallite transport properties. This transport is electronic in nature since (1) the corresponding thermal activation energy is fairly low (≈ 1 eV), and (2) the capacitance is constant over the frequency range considered (10^{-3}–100 Hz). The number of charge carriers per unit volume can be estimated from $\sigma = ne\mu$, assuming that the mobility is between 10^{-6} and 10^{-2} cm^2/V s. By taking the highest value of the mobility, the concentration of charge carriers is more than 10^{11} carriers /cm^3 for $[(C_{12}O)_8Pc]_2Lu$ and $(C_{12}O)_8PcH_2$, more than 10^{10} carriers/cm^3 for $(C_{12}O)_8PcLi$.

At higher frequency (>10^2–10^5 Hz) the $\omega^{0.5}$ and $\omega^{0.8}$ dependencies are observed. The mechanism is probably associated with intracolumnar or intercolumnar processes. The ionization potentials and electron affinities of paraffinic and aromatic moieties are significantly different [158–160], and charge carriers must be localized on the aromatic subunits. All transport processes therefore involve the aromatic cores of the columns. It is possible to estimate the respective contributions of the intracolumnar and intercolumnar processes. The hopping probability is related to the orbital overlap J by the equation [161] :

$$W_H = v \, j \exp(-\Delta E / kT) \qquad (39)$$

where $v_j = J/2h$ and $J = J_0 \exp(-\alpha/r_0)$, ΔE is the barrier height, α the width of the intermolecular potential barrier, r_0 the effective electron radius, estimated at 2Å [162], J the exchange energy between the hopping sites, J_0 effective exchange interaction.

By taking the distances determined by x-ray diffraction, it can be estimated that intracolumnar hopping probability is 10^7 times larger than the intercolumnar one. ac electrical conductivities therefore mainly arise from intracolumnar processes at high frequencies. Recent studies on $[(C_{12}O)_8Pc]_2Lu$ could determine the dielectric anisotropy of partially oriented mesophases [174].

It has been found that two different frequency dependencies ($\omega^{0.8}$; $\omega^{0.5}$) are clearly distinguishable in the high frequency range, depending on the compound studied. This difference seems to be related to the degree of order within the columns. $(C_{12}O)_8PcH_2$ and $(C_{12}O)_8PcLi$ form ordered columnar liquid crystals: In both cases a $\omega^{0.5}$ dependency is obtained. $[(C_{18}OCH_2)_8Pc]_2Lu$ and $(C_{18}OCH_2)PcH_2$ show a $\omega^{0.8}$ dependency and form disordered columnar liquid crystals.

In the liquid crystalline domain of $[(C_{18}OCH_2)_8Pc]_2Lu$, the conductivity at 10 GHz is equal to $3.9 \times 10^{-5} \, \Omega^{-1} \, cm^{-1}$, very close to the value extrapolated from the frequency range 10^2–10^5 Hz, assuming a $\omega^{0.8}$ dependency. The same frequency dependence is therefore found from 10^2 to 10^{10} Hz. At 10^{10} Hz, the charge carrier cannot migrate over more than a few angstroms; this confirms the intracolumnar nature of the electrical conduction. By assuming an intracolumnar mobility in the range 10^{-4}–$10^{-2} \, cm^2/V.s$, an external applied field of 20 V/cm and a column length around 500 Å, it may be readily calculated that the intracolumnar conduction process should be detected for frequencies higher than 10^2–10^4 Hz. This value is in good agreement with the experimental observations. The thermal activation energies of conduction in the mesophases, and in the frequency range where intracolumnar processes are effective, vary from 0.1 to 1 eV. No apparent correlation with the energy needed to generate intrinsic charge carriers can be found. The thermal variation is therefore associated with some detrapping process of charge carriers.

In the isotropic phase of $[(C_{18}OCH_2)_8Pc]_2Lu$, a ω^0 dependency is observed in the frequency range 10^{-3}–10^3 Hz with an activation energy of 0.8 eV [163]. Capacitance measurements indicate that ionic conduction probably occurs. By assuming that the ion migrates over the interelectrode distance (20 μm), a mobility of 4×10^{-4} cm^2/V s may be calculated (field, 50 V/cm). Correspondingly, a diffusion coefficient D of 10^{-5} cm^2/s may be obtained from the equation :

$$\mu = eD/kT \qquad (40)$$

This value is in agreement with conventional ionic mobilities in solutions or in liquid crystals [164].

The electrical properties of various substituted metallo-phthalocyanines have been measured at 10 GHz as a function of temperature (Fig. 53) [165].

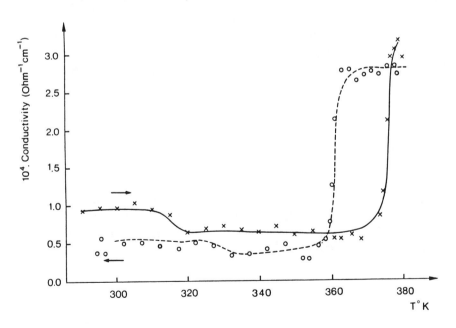

Figure 53 Electrical properties at 10 GHz of $(C_{12}O)_8PcLi$. The transition temperature to the mesophase is 365 K.

In all cases, the crystal to mesophase transition corresponds to an increase by a factor of 10 of the ac conductivity. The conductivity jump is reversible, and no decomposition occurs in the liquid crystalline phase. This enhancement of the conductivity probably corresponds to an increase of the mobility of the charge carriers.

K. CONCLUSION : TOWARDS SUBMICRONIC DEVICES

It has been demonstrated that columnar mesophases of substituted phthalocyanine derivatives may be used to induce one-dimensional electron or energy migrations. These materials may therefore be considered as submicronic [166] wires in which the active core, whose size is approximately 15 Å, is constituted of the phthalocyanine macrocycles. The active parts of the wires are isolated from each other by the paraffinic chains, which play the role of the insulator (the submicronic sheath of the cable). In standard computers, 98% of the space is devoted to connecting links. It is therefore of some importance to design wires at the submicronic scale. This is the first step towards submicronic systems, which should integrate the corresponding devices. It is not (too) difficult to design devices (i.e., field-effect transistors) that would take advantage of the columnar organization described in the previous chapters; one hopes that the realization of such devices will be the object of future work.

ACKNOWLEDGMENT

Nous voulons exprimer notre gratitude à toutes les personnes qui participèrent aux travaux décrits précedemment. Quelques unes ont pu ainsi rêver. (D'ailleurs un rêve, mon amie, n'est pas une espérance: on s'en contente... M. Yourcenar.) Ce travail a été effectué dans le Groupe de Recherches sur les Matériaux Moléculaires (GRIMM).

REFERENCES

1.　J. Simon, J.-J. André and A. Skoulios, *Nouv. J. Chimie* 10 (1986) 295.

2.　J. Simon, F. Tournilhac and J.-J. André, *New. J. Chem.* 11 (1987) 383.

3.　R. Feyman, in *La Nature de la physique*, Editions du Seuil, Paris, 1979.

4.　J. Simon, P. Bassoul and S. Norvez, *New J. of Chem.*, 13 (1989) 13.

5.　A. I. Kitaigorodskii, in *Organic Chemical Crystallography*, Consultants Bureau, New York, 1961 ; *Molecular Crystals and Molecules*, Academic Press, New York, 1973.

6.　A. Gavezotti, *J. Am. Chem. Soc.*, 105 (1983) 5220.

7.　M. L. Connolly, *J. Am. Chem. Soc.* 107 (1985) 1118.

8.　P. Kollman, *Acc. Chem. Res.*, 18 (1985) 105.

9.　B. Grünbaum and G. C. Shepard, in *Tilings and Patterns*, W. H. Freeman and Co. New York, 1987)

10.　M. Gardner, *Scientific American*, 233 (1975) 112.

11.　P. Bassoul and J. Simon, to be published.

12.　C. Piechocki, J. Simon, A. Skoulios, D. Guillon and P. Weber, *J. Am. Chem. Soc.*, 104 (1982) 5245.

13.　C. Piechocki, and J. Simon, *Nouv. J. Chimie*, 9 (1985) 159.

14.　M. J. Cook, M. F. Daniel, K. J. Harrison, N. V. Mc Keown, and A. J. Thomson, *J. Chem. Soc. Chem. Comm.* (1987) 1086 ; Materials 1 (1989) 287.

15.　M. Hanack, A. Beck and H. Lehman, *Synthesis* (1987) 703.

16.　B. A. Gregg, M. A. Fox and A. J. Bard, *J. Chem. Soc. Chem. Comm.* (1987) 1134.

17.　C. Tanford, *Science*, 200 (1978) 1012.

18.　A. Skoulios and G. Finoaz, *J. Chim. Phys.*, 59 (1962) 473.

19.　J. H. Hildebrand, and R. L. Scott, in *Regular Solutions*, Prentice Hall, Inc., Englewood Cliffs, 1962.

20.　*Polymer Handbook*, J. Brandrup, E. H. Immergut, eds., J. Wiley & Sons, New York, 1975.

21.　P. A. Small, *J. Appl. Chem*, 3 (1953) 71 (cited in [20])

22.　K. L. Hoy, *J. Paint Technol.*, 42 (1970) 76 (cited in [20]).

23.　C. Tanford, *J. Phys. Chem.* 78 (1974) 2469.

24.　A. Skoulios, *Adv. Colloid. Int. Sci.*, 1 (1967) 79.

25.　C. Piechocki and J. Simon, *J. C. S. Chem. Commun.* (1985) 259.

26.　S. Gaspard, P. Maillard and J. Billard, *Mol. Cryst. Liq. Cryst.* , 123 (1985) 369.

27.　The same effect is observed for nickel dithiolene complexes.

28.　L. Pauling, in *The Nature of the Chemical Bond*, third ed., Cornell University Press., New York, 1962.

29. P. G. de Gennes, *J. Phys. Lett.* 441 (1983) L-657.

30. C. Godreche and L. de Seze, *J. Phys. Lett.*, 46 (1985) L-39.

31. C. Sirlin, L. Bosio, J. Simon, V. Ahsen, E. Yilmazer and O. Bekaroglu, *Chem. Phys. Lett.*, 139 (1987) 362.

32. D. Andelman and P. G. de Gennes, *C. R. Acad. Sci. Paris,* 307 (1988) 233.

33. L. Edwards and M. Gouterman, *J. Mol. Spectros.*, 33 (1970) 292.

34. I. Chen, *J. Mol. Spectra*, 23 (1967) 131.

35. E. Canadell and S. Alvarez, *Inorg. Chem.*, 23 (1984) 573.

36. W. J. Pietro, T. J. Marks and M. A. Ratner, *J. Amer. Chem. Soc.*, 107 (1985) 5387.

37. M. Le Contellec, B. Vineuze, F. Richou, J. L. Favennec and J. Herrou, *Eurodisplay* (1984).

38. R. S. Mc Ewen, *J. Phys. E*, 20 (1987) 364.

39. I. A. Shanks, *Contemp. Phys.*, 23 (1982) 65.

40. J. Simon and C. Sirlin, *Pure Appl. Chem.*, 61 (1989) 1625.

41. W. J. Young, I. Haller and D. C. Green, *Mol. Cryst. Liq. Cryst.*, 13 (1971) 305.

42. W. R. Krigbaum, J. C. Poirier and M. J. Costello, *Mol. Cryst. Liq. Cryst.*, 20 (1973) 133.

43. J. Malthete and J. Billard, *Mol. Cryst. Liq. Cryst.*, 34 (1976) 117.

44. L. Verbit and T. R. Halbert, Mol. *Cryst. Liq. Cryst.*, 30 (1975) 209.

45. J. M. Wilson, R. Harden and J. Phillips, *Mol. Cryst. Liq. Cryst.*, 34 (1977) 237.

46. A.-M. Giroud and U. T. Mueller-Westerhoff, *Mol. Cryst. Liq. Cryst.*, 41 (1977) 11.

47. B. J. Bulkin, R. K. Mose and A. Santoro, *Mol. Cryst. Liq Cryst.*, 43 (1977) 53.

48. J. Le Moigne, Ph. Gramain and J. Simon, *J. Colloïd. Int.*, 60 (1977) 565.

49. J. Le Moigne and J. Simon, *J. Phys. Chem.*, 84 (1980) 170.

50. D. Markovitsi, A. Mathis, J. Simon, J.-C. Wittmann and J. Le Moigne, *Mol. Cryst. Liq. Cryst.*, 64 (1980) 121.

51. A. M. Giroud, A. Nazzal and U. T. Mueller-Westerhoff, *Mol. Cryst. Liq. Cryst.*, 56 (1980) 225.

52. U. T. Mueller-Westerhoff, A. Nazzal, R. J. Cox and A. M. Giroud, *J. C. S. Chem. Comm.* (1980) 497.

53. M. Cotrait, J. Gaultier, C. Polycarpe and A. M. Giroud, *Acta. Cryst.*, C39 (1983) 833.

54. K. Ohta, M. Yokoyama, S. Kusabayashi and H. Mikawa, *J. C. S. Chem. Comm.* (1980) 392.

55. M. Ghedini, M. Longeri and R. Bartolino, *Mol. Cryst. Liq. Cryst.*, 84 (1982) 207.

56. M. Veber, R. Fugnito and H. Strzelecka, *Mol. Cryst. Liq. Cryst.*, 96 (1983) 221.

57. A.-M. Giroud-Godquin, and J. Billard, *Mol. Cryst. Liq. Cryst.*, 66 (1981) 147; 97 (1983) 287.

58. A.-M. Giroud-Godquin, G. Sigaud, M. F. Achard and F. Hardouin, *J. Phys. Lett.*, 45 (1984) L387.

59. H. Sakashita, A. Nishitani, Y. Sumiga, H. Terauchi, K. Ohta and I. Yamamoto, *Mol. Cryst. Liq. Cryst.*, 163 (1988) 211.

60. K. Ohta, H.Muroki, A. Takagi, K.-I. Matada, H. Ema, I. Yamamoto and K. Matzuraki, *Mol. Cryst. Liq. Cryst.*, 140 (1986) 131.

61. B. K. Sadashiva and S. Ramesha, *Mol. Cryst. Liq. Cryst.*, 141 (1986) 19.

62. S. Chandrasekhar, B. K. Sadashiva and B. S. Srikanta, *Fifth European Winter Liquid Crystal Conference*, Borovets, Bulgaria, March, 1987.

63. S. Gaspard, A. Hochapfel and R. Viovy, *C. R. Acad. Sci.*, Paris, 289 ser. C (1979) 387.

64. A. Skoulios, personal communication.

65. J. W. Goodby, P. S. Robinson, Boon-Keng Teo and P. E. Cladis, *Mol. Cryst. Liq. Cryst.*, 56 (1980) 303.

66. A. B. P. Lever, *Adv. Inorg. Chem. and Radiochem.*, 7 (1965) 27.

67. J. Simon and J.-J. André, in *Molecular Semiconductors*, Springer Verlag, Berlin, 1985.

68. G. Pawlowski and M. Hanack, *Synthesis* (1980) 287.

69. E. Orthman and G. Wegner, *Angew. Chem.*, Int. ed. Engl., 25 (1986) 1105.

70. J. Metz, O. Schneider and M. Hanack, *Inorg. Chem.*, 23 (1984) 1065.

71. H. Meier, W. Albrecht, D. Wöhrle and A. Jahn, *J. Phys. Chem.*, 90 (1986) 6349.

72. M. J. Camenzind and C. L. Hill, *J. Heterocycl. Chem.*, 22 (1985) 575.

73. B. Pawlowski and M. Hanack, *Synth. Comm.*, 11 (1981) 351.

74. K. Ohta, L. Jacquemin, C. Sirlin, L. Bosio and J. Simon, *New J. Chem.*, 12 (1988) 751.

75. D. Masurel, C. Sirlin and J. Simon, *New J. Chem.*, 11 (1987) 455.

76. J. F. van der Pol, E. Neelemean, J. W. Zwikker, R. J. M. Nolte and W. Drenth, *Recl. Trav. Chim.*, Pays-Bas, 107 (1988) 615.

77. K. Tamao, K Sumitani and M. Kumada, *J. Amer. Chem. Soc.*, 94 (1972) 4374 ; *Org. Synth.*, 58 (1978) 127.

78. M. J. Cook, M. F. Daniel, K. J. Harrison, N. B. Mc Keown and A. J. Thomson, *J. C. S. Chem. Comm.* (1987) 1148.

79. N. Kobayashi, M. Koshiyama, Y. Ishikawa, T. Osa, H. Shirai and N. Hojo, *Chem. Letters* (1984) 1633.

80. H. Meyer and K. Steiner, *Monatsch. Chem.*, 35 (1914) 391.

81. E. A. Lawton and D. D. Mc Ritchie, *J. Org. Chem.*, 24 (1959) 26.

82. D. R. Boston and J. C. Bailer, Jr., *Inorg. Chem.*, 11 (1972) 1978.

83. P. A. Barrett, D. A. Frye and R. P. Linstead, *J. Chem. Soc.* (1938) 1157.

84. D. Wöhrle and G. Meyer, *Makromol. Chem.*, 181 (1980) 2127.

85. G. T. Byrne, R. P. Linstead and A. R. Lowe, *J. Chem. Soc.* (1934) 1017.

86. J. A. Elvidge and N. R. Barot, in *The Chemistry of Double Bonded Functional Group*, S. Patai, ed., J. Wiley and Sons, New York, 1977.

87. P. J. Brach, S. J. Grammatica, O. A. Ossanna and L. Weinberger, *J. Heterocycl. Chem.*, 7 (1970) 1403.

88. C. Piechocki, *Thèse de Doctorat d'Etat*, ESPCI Paris, France, 1985.

89. D. Lelièvre, J. Simon and M Petit, *Liq. Cryst.*, 4 (1989) 707.

90. D. Demus and L. Richter, in *Textures of Liquid Crystals*, Verlag Chemie, Weinheim, 1978.

91. S. Chandrasekhar, B. K. Sadashiva and K A. Suresh, *Pramana* (1977) 471.

92. *CRC Handbook of Chemistry and Physics*, R. C Weast and M. J. Astle, ed., CRC, Boca Raton, 1980.

93 D. Guillon, A. Skoulios, C. Piechocki, J. Simon and P. Weber, *Mol. Cryst. Liq. Cryst.*, 100 (1983) 275.

94. P. Seurin, D. Guillon and A. Skoulios, *Mol. Cryst. Liq. Cryst.*, 65 (1981) 85.

95. A. M. Levelut, *J. Phys. Lett.*, 40 (1979) 81.

96. D. Guillon, P. Weber, A. Skoulios, C. Piechocki and J. Simon, *Mol. Cryst. Liq. Cryst.*, 130 (1985) 223.

97. A. M. Levelut, *J. Chim. Phys.*, 80 (1983) 149.

98. C. Destrade, Nguyen Hun Tinh, H. Gasparoux, H. Malthete and A. M. Levelut, *Mol. Cryst. Liq. Cryst.*, 71 (1981) 111.

99. C. C. Leznoff and Tse Wai Hall, *Tetrahedron Lett.*, 23 (1982) 3023.

100. C. C. Leznoff, *Nouv. J. de Chim.*, 6 (1982) 653.

101. V. I. Minkin, O. A. Osipov and Y. A. Zhdavov, in *Dipole Moments in Organic Chemistry*, Plenum Press, New York, 1970.

102. T. J. Janson, A. R. Kane, J.-F. Sullivan, K. Knox and M. E. Kenney, *J. Amer. Chem.*, 91 (1969) 5210.

103. C. J. F. Bötcher, in *Theory of Electric Polarization*, Elsevier, Amsterdam, 1973.

104. C. Piechocki, J. C. Boulou and J. Simon, *Mol. Cryst. Liq. Cryst.*, 149 (1987) 115.

105. P. Weber, D. Guillon and A. Skoulios, *J. Phys. Chem.*, 91 (1987) 2242.

106. K. Ukei, *Acta. Cryt.*, B29 (1973) 2290.

107. M. K. Friedel, B. F. Hoskins, R. L. Martin and S. A. Mason, *J. C. S. Chem. Comm.* (1970) 400.

108. U.Iyechika, K. Yakushi, I Ikemoto,and H. Kuroda, *Acta. Cryst.*, B28 (1982) 766.

109. F. Przyborowski and C. Hamann, *Crystals Res. and Technol.*, 17 (1982) 1041.

110. Th. Frauenheim, C. Hamann and M. Müller, *Phys. Status Solidi*, 86 (1984) 735.

111. C. W. Dirk, T. Inabe, K. F. Schoch, Jr. and T. J. Marks, *J. Amer. Chem. Soc.*, 105 (1983) 1539.

112. C. Sirlin, L. Bosio and J. Simon, *J. C.S. Chem. Comm.* (1987) 379.

113. W. J. Kroenke, L. E. Sutton, R. D. Joyner and M. E. Kenney, *Inorg. Chem.*, 2 (1963) 1064 ; 1 (1962) 717.

114. J. Metz, G. Pawlowski and M. Hanack, *Z. Naturforschung*, Teil B, 38 (1983) 378;

115. R. D. Joyner and M. E. Kenney, *Inorg. Chem.*, 1 (1962) 236.

116. M. K. Lowery, A. J. Starshak, J. N. Esposito, P. C. Krueger and M. E. Kenney, *Inorg. Chem.*, 3 (1964) 128.

117. P. M. Kuznesof, R. S. Nohr, K. J. Wynne and M. E. Kenney, *J. Macromol. Sci. Chem.*, 16 (1981) 299.

118. C. Sirlin, L. Bosio and J. Simon, *J. C. S. Chem. Comm.*, 236 (1988); *Mol. Cryst. Liq. Cryst.*, 155 (1988) 231.

119. A. R. Kane, J. F. Sullivan, D. H. Kenney and M. E. Kenney, *Inorg. Chem.*, 9 (1970) 1445.

120. J. Friedrich and D. Haarer, *Angew. Chem. Int. ed.*, 23 (1984) 113.

121. L. A. Rebane, A. A. Gorokhovskii and J. V. Kikass, *App. Phys.*, B29 (1982) 235.

122. S. Völker and J. H. van der Waals, *Mol. Physics*, 32 (1976) 1703.

123. J. McVie, R. S. Sinclair and T. G. Truscott, *J. Chem. Soc. Faraday Trans II*, 74 (1978) 1870.

124. R. Knoesel, C. Piechocki and J. Simon, *J. Photochem*, 29 (1985) 445.

125. W. F. Kosonocky and S. E. Harrison, *J. App. Phys.*, 37 (1966) 4789.

126. K. A. Zachariasse and D. G. Whitten, *Chem. Phys. Letters*, 86 (1982) 228.

127. W. E. Blumberg and J.Peisach, *J. Biol. Chem.*, 240 (1965) 870.

128. J.-J. André, M. Bernard, C. Piechocki and J. Simon, *J. Phys. Chem.*, 90 (1986) 1327.

129. K. Ito and Y. Yamashita, *J. Polym. Sci.* Part. A3 (1965) 2165.

130. B. Blanzat, C. Barthou, N. Tercier, J.-J. André and J. Simon, *J. Amer. Chem. Soc.*, 109 (1987) 6193.

131. R. O. Loutfy, J. H. Sharp, C. K. Hsiao and R. J. Ho, *J. App. Phys.*, 52 (1981) 5218.

132. D. Markovitsi, Thu-Hoa Tran-Thi, V. Briois, J. Simon and K. Ohta, *J. Amer. Chem. Soc.*, 110 (1988) 2001.

133. P. Jacques and A. M. Braun, *Helv. Chim. Acta.*, 64 (1981) 1800.

134. D. Markovitsi, I. Lecuyer and J. Simon, *J. Phys. Chem.*, 95 (1991) 3620.

135. L. E. Lyons, *Aust. J. Chem.*, 33 (1980) 1717.

136. J. Simon, *l'Actualité Chimique*, 349 Nov–Déc. 1987.

137. J. Simon, J.-J. André and M. Maitrot in *Molecules in Physics, Chemistry and Biology*, J. Maruani, ed., D. Reidel, Dordrecht, 1988.

138. P. Turek, P. Petit, J.-J. André, J. Simon, R. Even, B. Boudjema, G. Guillaud and M. Maitrot, *J. Amer. Chem. Soc.*, 109 (1987) 5119.

139. J.-J. André, K. Holzcer, M. Petit, M.-T. Riou, C. Clarisse, R. Even, M. Fourmigué and J. Simon, *Chem. Phys. Letters*, 115 (1985) 463.

140. M. Maitrot, G. Guillaud, B. Boudjema, J.-J. André, H. Strzelecka, J. Simon and R. Even, *Chem. Phys. Letters*, 133 (1987) 59.

141. P. N. Moskalev and I. S. Kirin, *Russ. J. Inorg. Chem.*, 16 (1971) 57.

142. C. Piechocki, J. Simon, J.-J. André, D. Guillon, P. Petit, A. Skoulios and P. Weber, *Chem. Phys. Letters*, 122 (1985) 124.

143. Z. Belarbi, C. Sirlin, J. Simon and J.-J. André, *J. Phys. Chem.*, 93 (1989) 8105.

144. A. T. Chang and J. C. Marchon, *Inorg. Chimica Acta*, 53 (1981) L241.

145. J. C. Marchon, *J. Electrochem. Soc.*, 129 (1982) 1377.

146. G. A. Corker, B. Grant and N. J. Clecak, *J. Electrochem. Soc.*, 126 (1979) 1339.

147. P. N. Moskalev and I. S. Kirin, *Opt. Spectrosc.*, 29 (1970) 220.

148. G. C. S. Collins and D. J. Schiffrin, *J. Electronal. Chem.*, 139 (1982) 335.

149. M. M. Nicholson and F. A. Pizzarello, *J. Electrochem. Soc.*, 126 (1979) 1490.

150. M. M'Sadak, J. Roncali and F. Garnier, *J. Electroanal. Chem.*, 189 (1985) 99.

151. M. T. Riou, M. Auregan and C. Clarisse, *J. Electroanal. Chem.*, 187 (1985) 349.

152. M. L'Her, Y. Cozien and J. Courtot-Coupez, *C. R. Acad. Sci.*, Paris, 300 (1985) C487; J. Electroanal. Chem., 157 (1983) 183.

153. F. Castaneda, C. Piechocki, V. Plichon, J. Simon and J. Vaxivière, *Electrochimica Acta.*, 31 (1986) 131.

154. R. M. Schaffert, in *Electrophotography Focal*, Press London, 4th edition, 1975.

155. S. Besbes, V. Plichon, J. Simon and J. Vaxivière, *J. Electroanal. Chem.*, 237 (1987) 61.

156. E. Markovitsi, Thu-Hoa Tran-Thi, R. Even and J. Simon, *Chem. Phys. Letters*, 137 (1987) 107.

157. J. C. Dyre, *J. App. Phys.*, 64 (1988) 2456.

158. J. Partridge, *J. Chem. Phys.*, 45 (1966) 1679.

159. W. L. Mc Cubbin and I. D. C. Gurney, *J. Chem. Phys.*, 43 (1965) 983.

160. J. Mort and D. M. Pai in *Photoconductivity and Related Phenomena*, Elsevier, Amsterdam, 1976.

161. H. Meier, in *Organic Semiconductors*, Verlag Chemie, Weinheim, 1974.

162. M. Stolka, J. F. Yanus and D. M. Pai, *J. Phys. Chem.*, 88 (1984) 4707.

163. Z. Belarbi, M. Maitrot, K. Ohta, J. Simon, J.-J. André and P. Petit, *Chem. Phys. Letters*, 143 (1988) 400.

164. M. Yamashita and Y. Amemiya, *Jpn. J. App. Phys.*, 17 (1978) 1513.

165. J.-J. André, P. Petit and J. Simon, to be published.

166. *Submicron* was the term given by Zsigmondy (1865–1929) and Siedentopf (1872–1940) to colloidal particles that are too small to be observed by optical microscopy.

167. P. G. Schouten, J. F. van der Pol, J. W. Zwikker, W. Drenth and S. J. Picken, *Mol. Cryst. Liq. Cryst.* 195 (1991) 291.

168. P. Weber, D. Guillon and A. Skoulios, *Liq. Crystals* 9 (1991) 369.

169. A.-M. Giroud-Godquin and P. M. Maitlis, *Angew. Chem. Int.* ed Engl. 3O(1991) 375.

170. K. Ohta, T. Watanabe, T. Fujimoto and I. Yamamoto, *J. C. S. Chem. Comm.* (1989) 1611.

171. K. Ohta, T. Watanabe, H. Hasebe, Y. Morizumi, T. Fujimoto, I. Yamamoto, D. Lelièvre and J. Simon, *Mol. Cryst. Liq. Cryst.* 196 (1991) 13.

172. I. Cho and Y.-S. Lim, *Bull. Korean Chem. Soc.* 9 (1988) 98.

173. C. C. Leznoff and A. B. P. Lever Ed. in *Phthalocyanines,* Vol I, VCH, Weinheim, 1989.

174. Z. Belarbi, *J. Phys. Chem.,* 94 (1990) 7334.

Index